Sustainable Civil Infrastructures

About this Series

Sustainable Infrastructure impacts our well-being and day-to-day lives. The infrastructures we are building today will shape our lives tomorrow. The complex and diverse nature of the impacts due to weather extremes on transportation and civil infrastructures can be seen in our roadways, bridges, and buildings. Extreme summer temperatures, droughts, flash floods, and rising numbers of freeze-thaw cycles pose challenges for civil infrastructure and can endanger public safety. We constantly hear how civil infrastructures need constant attention, preservation, and upgrading. Such improvements and developments would obviously benefit from our desired book series that provide sustainable engineering materials and designs. The economic impact is huge and much research has been conducted worldwide. The future holds many opportunities, not only for researchers in a given country, but also for the worldwide field engineers who apply and implement these technologies. We believe that no approach can succeed if it does not unite the efforts of various engineering disciplines from all over the world under one umbrella to offer a beacon of modern solutions to the global infrastructure. Experts from the various engineering disciplines around the globe will participate in this series, including: Geotechnical, Geological, Geoscience, Petroleum, Structural, Transportation, Bridge, Infrastructure, Energy, Architectural, Chemical and Materials, and other related Engineering disciplines.

More information about this series at http://www.springer.com/series/15140

Tarek Abdoun · Sherif Elfass
Editors

Soil Dynamics and Soil-Structure Interaction for Resilient Infrastructure

Proceedings of the 1st GeoMEast International Congress and Exhibition, Egypt 2017 on Sustainable Civil Infrastructures

Springer

2017 GEOMEAST

SSIGE
SOIL-STRUCTURE INTERACTION GROUP IN EGYPT

Editors
Tarek Abdoun
Rensselaer Polytechnic Institute (RPI)
Rensselaer, NY
USA

Sherif Elfass
University of Nevada, Reno
Reno, NV
USA

ISSN 2366-3405 ISSN 2366-3413 (electronic)
Sustainable Civil Infrastructures
ISBN 978-3-319-63542-2 ISBN 978-3-319-63543-9 (eBook)
DOI 10.1007/978-3-319-63543-9

Library of Congress Control Number: 2017946437

Printed on acid-free paper

This Springer imprint is published by Springer Nature
The registered company is Springer International Publishing AG
The registered company address is: Gewerbestrasse 11, 6330 Cham, Switzerland

Preface

Toward building sustainable and longer civil infrastructures, the engineering community around the globe continues undertaking research and development to improve existing design, modeling, and analytical capability. Such initiatives are also the core mission of the Soil-Structure Interaction Group in Egypt (SSIGE) to contribute to the ongoing research toward sustainable infrastructure. This conference series "GeoMEast International Congress and Exhibition" is one of these initiatives.

Ancient peoples built their structures to withstand the test of time. If we think in the same way, our current projects will be a heritage for future generations. In this context, an urgent need has quickly motivated the SSIGE and its friends around the globe to start a new congress series that can bring together researchers and practitioners to pursue "Sustainable Civil Infrastructures." The GeoMEast 2017 is a unique forum in the Middle East and Africa that transfers from the innovation in research into the practical wisdom to serve directly the practitioners of the industry.

More than eight hundred abstracts were received for the first edition of this conference series "GeoMEast 2017" in response to the Call for Papers. The abstracts were reviewed by the Organizing and Scientific Committees. All papers were reviewed following the same procedure and at the same technical standards of practice of the TRB, ASCE, ICE, ISSMGE, IGS, IAEG, DFI, ISAP, ISCP, ITA, ISHMII, PDCA, IUGS, ICC, and other professional organizations who have supported the technical program of the GeoMEast 2017. All papers received a minimum of two full reviews coordinated by various tracks chairs and supervised by the volumes editors through the Editorial Manager of the SUCI "Sustainable Civil Infrastructure" book series. As a result, 15 volumes have been formed of the final +320 accepted papers. The authors of the accepted papers have addressed all the comments of the reviewers to the satisfaction of the tracks chairs, the volumes editors, and the proceedings editor. It is hoped that readers of this proceedings of the GeoMEast 2017 will be stimulated and inspired by the wide range of papers written by a distinguished group of national and international authors.

Publication of this quality of technical papers would not have been possible without the dedication and professionalism of the anonymous papers reviewers. The names of these reviewers appear in the acknowledgment that follows. For any additional reviewers whose names were inadvertently missed, we offer our sincere apologies.

We are thankful to Dr. Hany Farouk Shehata, Dr. Nabil Khelifi, Dr. Khalid M. ElZahaby, Dr. Mohamed F. Shehata, and to all the distinguished volumes editors of the proceedings of the GeoMEast 2017. Appreciation is extended to the authors and track chairs for their significant contributions. Thanks are also extended to Springer for their coordination and enthusiastic support to this conference. The editors acknowledge the assistance of Ms. Janet Sterritt-Brunner, Mr. Arulmurugan Venkatasalam in the final production of the 15 edited volumes "Proceedings of GeoMEast 2017".

Contents

The Deflection and Bending Moment of Existing Piles Influenced by Trenching Diaphragm Wall Panel(s)

Ahmed Mohamed[1(✉)], Marawan Shahin[2], and Herbert Klapperich[1]

[1] Geo-Institute, TU Bergakademie Freiberg, Freiberg, Germany
ahmed.mohamed01@feng.bu.edu.eg, hdkl111@t-online.de
[2] Faculty of Engineering, Tanta University, Tanta, Egypt
marawan.shahin@feng.tanta.edu.eg

Abstract. Diaphragm walls are normally used for the deep excavations and underground construction inside the cities. The trenching process of diaphragm walls causes deformation in the ground surface. The construction of the diaphragm walls near deep foundations may not be avoided specially in the crowded cities. This research focuses on the deflection and bending moment of piled foundations near slurry trench. Since the trench is considered as a three dimensional problem, a three-dimensional numerical analysis was used in this research. It was conducted using a commercial analysis software known as FLAC 3D which is based on the finite different analysis method. Verification was made for the numerical analysis method using two case histories, one in Hong-Kong and another one in Giza, Egypt. The results from the numerical analysis were in quite a good contrast with the field data results. This means the trenching process could be numerically modeled with the proposed method and provides good results. The verified numerical analysis method was used to conduct a numerical parametric study that discusses the effect of a single or double panel(s) on the nearby piled foundation. The parametric study showed that the pile deflection and bending moment are affected by the panel length, pile diameter, pile location from the trench, groundwater level and the existence of a weak soil layer that could cause reduction in slurry pressure. The panel thickness did not show any influence on the pile bending moment or deflection. The decrease of the panel length and controlling the slurry level will reduce the influence of trenching on nearby piles. Lowering the groundwater level before trenching could probably reduce the influence on the pile as well.

Keywords: Diaphragm wall · Pile · Trenching · Deflection · Bending moment · Numerical analysis

1 Introduction

The soil deformation caused by the diaphragm wall trenching process was monitored by many researchers. Generally, they used the settlement points to measure the surface soil settlement and inclinometers to measure the horizontal soil displacement. Deep soil settlements could also have been measured in some cases. Monitoring during trenching

T. Abdoun and S. Elfass (eds.), *Soil Dynamics and Soil-Structure Interaction for Resilient Infrastructure*, Sustainable Civil Infrastructures, DOI 10.1007/978-3-319-63543-9_1

process was made for either tested panel or multiple panels. DiBiagio and Myrvoll (1972), Tsai et al. (2000) and Ng et al. (1999) had intensely monitored tested panel for research purposes, while Karlsrud (1983), Cowland and Thorley (1985), Hamza et al. (1999), Poh et al. (2001) and L'Amante et al. (2012) monitored the diaphragm wall installation of case histories in real projects. The monitoring results were varied according to many parameters such as panel dimensions, groundwater level and soil properties.

Trench panels were simulated using three-dimensional numerical analysis by Ng and Yan (1998), Gourvenec and Powrie (1999), Grandas-Tavera (2012) and Comodromos et al. (2013). These researchers compared the results from the numerical analysis with that from the field data. There intension was to find out the ability of the trenching numerical analysis in modeling the trenching process. They found out that three-dimensional numerical analysis could be used in modeling the trenching problem.

Generally, the trenching process causes settlement and horizontal displacement for the ground surface which could probably affect the nearby existing deep foundations. However, a very limited research has been made regarding such effect. Davies and Henkel (1982), Abdel-Rahman and El-Sayed (2009) and Korff (2013) monitored the trenching process near existing piled foundation but they were not able to monitor the existing piles. Choy et al. (2007) studied the effect of the trenching process on a single pile using the centrifuge model test through conducting a limited parametric study. His study did not take into consideration the effect on pile group and the possible existence of groundwater.

In this research, the three-dimensional numerical analysis was used to simulate the trenching process of diaphragm walls adjacent to piled foundation for two different case histories. The results from the numerical analysis were compared to the field data results. Such simulation method was used to conduct a parametric study which discusses the different parameters that affect the existing piled foundation near diaphragm wall panel(s).

2 Numerical Analysis of Case Histories and Verification

The numerical modeling for different geotechnical engineering problems is considered to be an acceptable tool. However, the modeling method of the different types of such problems should be verified. The trenching process of slurry trench walls required a three-dimensional simulation with a special attention to the stages and simulation assumptions. This section describes the numerical modeling of two case histories using FLAC 3D. The results from modeling were compared to those from the field.

2.1 Case History 1 (Underground Station Near the Court of Justice in Hong Kong)

A diaphragm wall system was used as a part of the Charter underground metro station. This station was very close to the court of justice which was constructed in a timber piled foundation. The piles cross section area was equal to 0.0254 m^2 and it extended to a level of 14 m beneath the ground surface. The building load was distributed over the beams that connect the piles. The details of the project are described by Davies and Henkel (1982). The numerical modeling and verification related to the diaphragm wall panels construction adjacent to the building are described in the following subsections.

2.1.1 Modeling and Construction Stages

The construction stages of the diaphragm wall panels were modeled as was described in Fig. 1. According to Stround and Sweeney (1977) the soil was found to be consisted of five layers. The soil layers' depths and their properties are presented in Table 1. The soil was modeled using strain hardening softening soil model which is defined in FLAC 3D by conducting a relation between mobilized friction angle and plastic shear strain which can be calculated according to Byrne (2003) as:

$$\xi_p = \frac{P_{ref}}{\beta G^e_{ref}} \times \frac{\sin \phi}{R_f}\left(\left(1 - \frac{\sin \phi_m}{\sin \phi} R_f\right)^{-1} - 1\right) \tag{1}$$

where

P_{ref} is the reference pressure,
β is calibration factor,
ϕ is the ultimate friction angle,
ϕ_m is the mobilized friction angle,
R_f is the failure ratio.

The elastic tangent shear modulus is calculated from the following equation:

$$G^e_{ref} = \frac{E^{ref}_{ur}}{2(1 + \upsilon_{ur})} \tag{2}$$

where

E^{ref}_{ur} is the required strain to mobilize the limit friction angle,
ν_{ur} is the undrained Poisson's ratio.

The relation between the plastic shear strain and mobilized friction angle based on the previous equations for the soil layers are presented in Fig. 2.

The trenching process of each panel was made by replacing the soil elements at the panel location with a hydrostatic pressure equivalent to the slurry pressure. The concreting process of the panel is made then by reactivation of the zones and changing its properties to concrete properties. Figure 3 shows the trenching process of a panel in different situations. The pile and beams connecting the piles were modeled using the beam and pile elements, respectively. These elements are described and discussed by Itasca (2012). The mesh model contains 157800 zones and its shape and dimension are presented in Fig. 4. The relative normal and shear displacement between the pile and

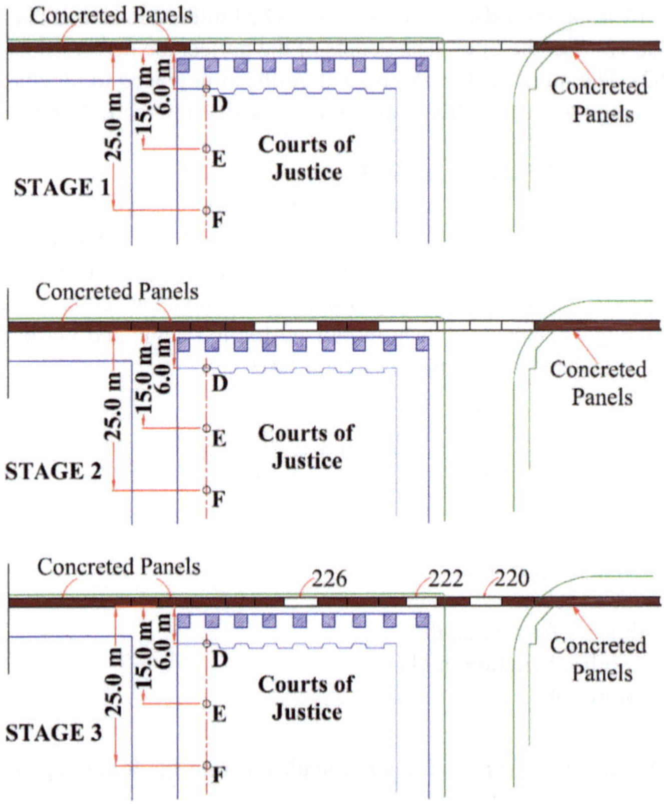

Fig. 1. Courts of justice construction stages and points of monitoring

Table 1. Soil properties (Courts of Justice, Hong Kong)

Soil layer	Bottom level (m)	SPT	γ (kN/m^3)	c′/cu (kN/m^2)	ϕ' (°)	E (MPa)
Fill	−1.0	5–30	17.0	0	27	10
Marine deposits	−2.6	5–30	17.0	5/35	15	10
Highly weathered Decomposed granite	−11.0	10–20	17.62	0	30	40
Decomposed granite	−31.75	>40	20.0	0	36	85
Granite	–	–	22.2	0	40	100

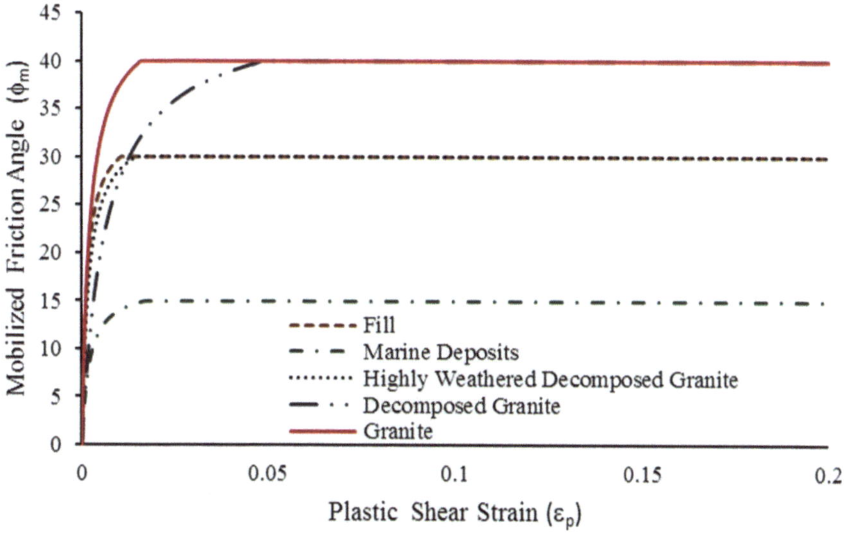

Fig. 2. Relation between mobilized friction angle and plastic shear strain for different soil layers

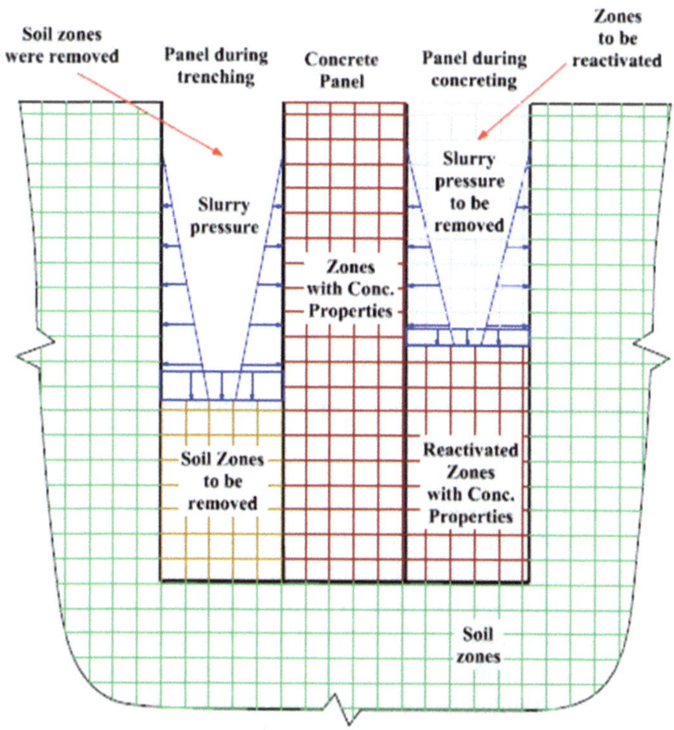

Fig. 3. Trench modeling process

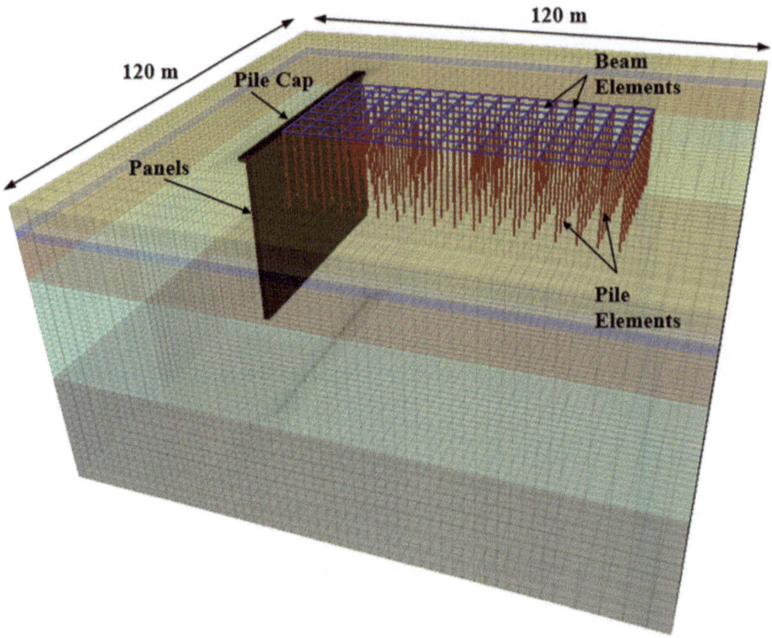

Fig. 4. Mesh Geometry of trench near court of justice

the soil are defined by the normal stiffness K_n and shear stiffness Ks, respectively. They are considered to be equal and can be calculated from the following empirical equation:

$$K_n = K_s = 10\max\left[K + \frac{4G}{\Delta z_{\min}}\right] \tag{3}$$

where

k is the soil bulk modulus

G is the soil shear modulus

Δz_{min} is the minimum distance in the vertical direction of the mesh

2.1.2 Results and Comparison

The settlement results of the numerical analysis compared to field data for points D and E at different stages are presented in Fig. 5. The results showed that the settlement values increase with stages and decrease with the distance from the trench.

The values of settlement from the numerical analysis nearest to the trench (point D) were in a very good agreement with those from the field data. This good agreement was not found in case of comparing the results at point E. Generally, the comparison showed that the numerical analysis provides reliable results that can present the reality.

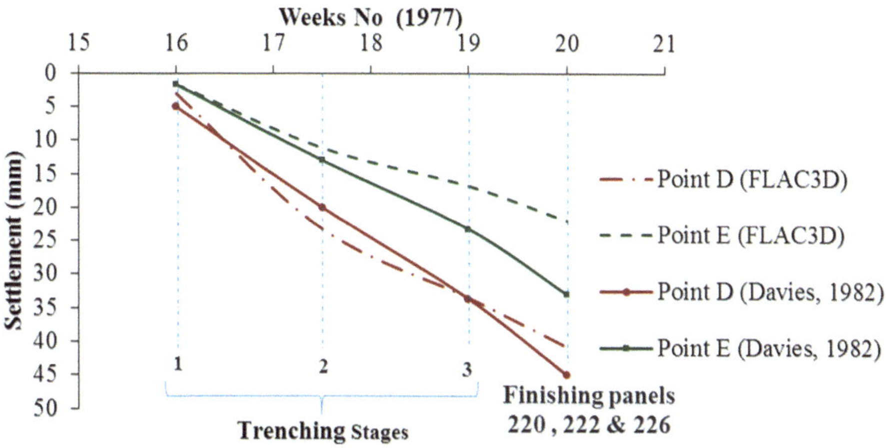

Fig. 5. Settlement at different stages

2.2 Case History 2 (Basement Near High-Rise Buildings in Giza, Egypt)

The underground construction of a basement in a crowded area in Giza, Egypt was done using diaphragm wall technique. The construction area was surrounded by several buildings as described by Abdel-Rahman and El-Sayed (2009). The soil was mainly sand and it was simulated using the same soil model that was used in the first case history. The soil properties are shown in Table 2. The panels construction stages are presented in Fig. 6 and they were modelled as in the field. Each panel was modeled as previously described and as shown in Fig. 3.

Table 2. Soil properties (Basement near piled foundation, Giza)

Soil layer	Bottom level (m)	SPT	γ_b (kN/m^2)	ϕ' (°)	E_{oed} (MPa)	E_{ur} (MPa)
Fill	−2.0	–	17.0	28	16.0	48.0
Silty sand	−5.0	12	18.0	30	17.0	49.0
Medium sand	−11.0	20	19.0	33.5	36.0	108.0
Dense sand	−25.0	42	20.0	36	42.0	124.0

The piles and the grade beams connecting the pile caps were modelled as described in the first case history. The pile caps were modeled using the shell elements and they carry the building load. The mesh used to simulate this case history is presented in Fig. 7.

The results from the numerical analysis compared to those from the field are presented in Fig. 8. The comparison was made for the three sections. The values of settlement were measured during trenching of the last panel (i.e. stage P-20-B) and it shows a decrease with distance from the trench. There was a slight difference regarding the settlement shape between the field and the numerical analysis. The settlement values from the numerical analysis adjacent to the trench and at a distance of 19 m

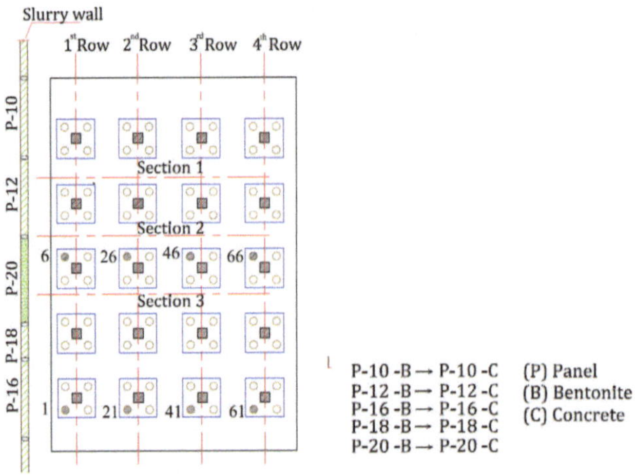

Fig. 6. The construction stages of the panels adjacent to the studied building

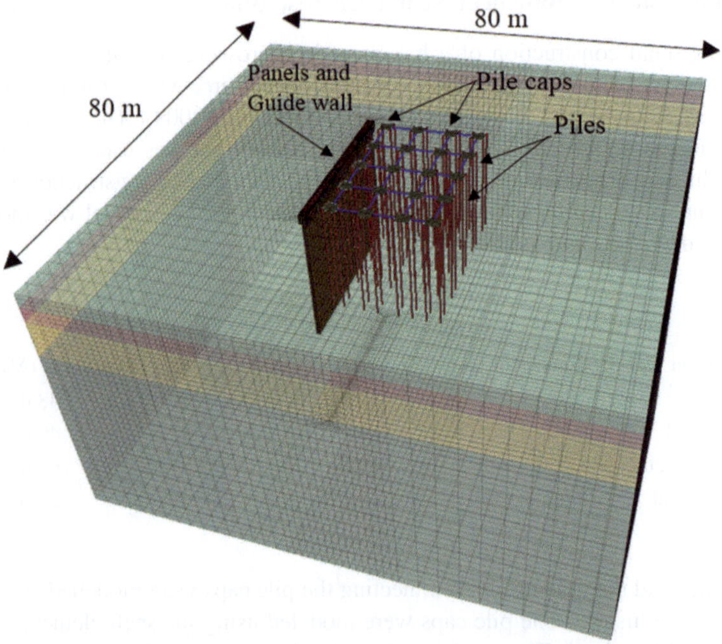

Fig. 7. Mesh geometry of the trench near the multi-story building

from the trench are identical with those from the field, but the settlement values from the numerical analysis were slightly higher than those from the field at a distance of 5 m from the trench. Generally, the output from the numerical analysis is in a good contrast with the field results.

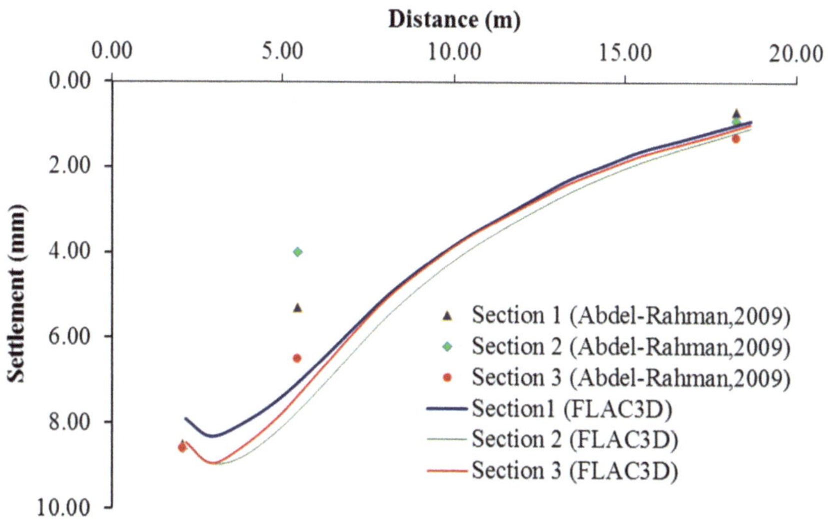

Fig. 8. Settlement at the last stage for different sections

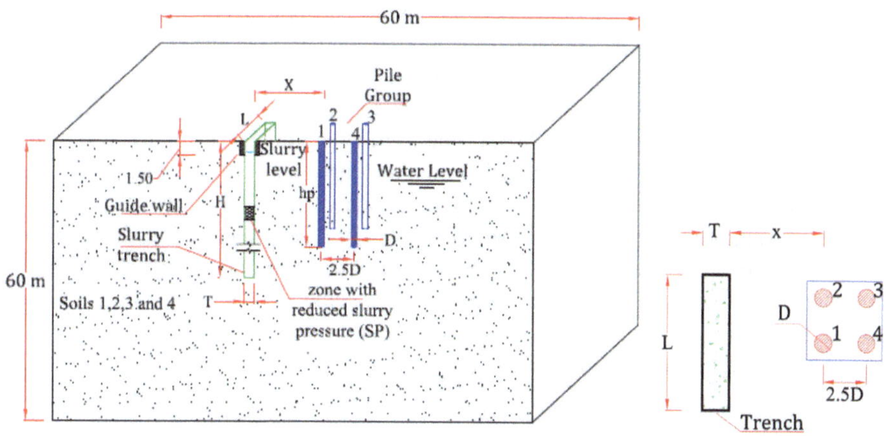

Fig. 9. Model group (MG1) – single panel

3 Numerical Parametric Study

The numerical simulation and comparison for the presented case histories showed that the method of simulating the trench numerically is acceptable and provides reliable results. This method was used to conduct a numerical parametric study includes several parameters. Two main models were used for the parametric study. The first model simulated the effect of the single panel on the pile group as presented in Fig. 9, while the second model simulated the effect of the double panels on similar pile group as shown in Fig. 10.

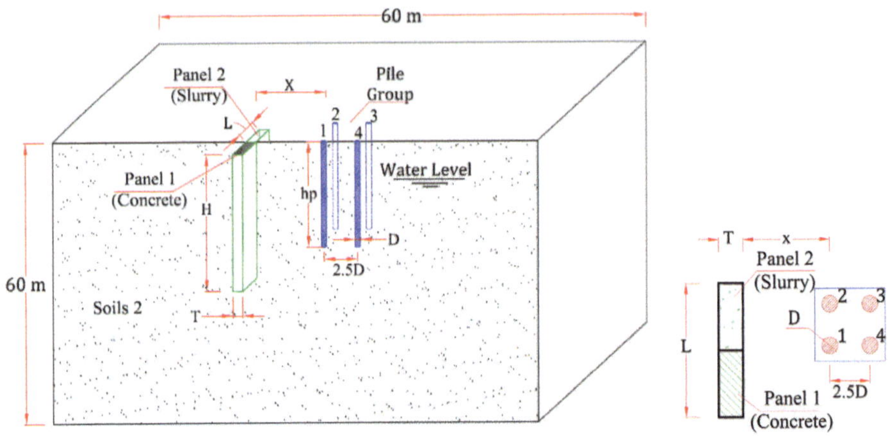

Fig. 10. Model group (MG2) – double panel

3.1 Studied Parameters

The soil used in this study was sand with a friction angle $\phi = 32°$, a bulk density $\gamma_b = 18$ kN/m^3 and shear modulus G = 9.6 MPa. The pile depth was 12 m and its diameter (D) was 0.8 m in all cases, but in case of studying the effect of the pile diameter and its properties were chosen as variables according to Table 3. The panel depth was chosen to be 30 m because below this depth the pile with a 12 m depth will almost not be affected (Mohamed 2015). The trench length (L) was either 3 m or 6 m while its thickness (T) was 0.6 m and 1.2 m. The slurry and groundwater levels were chosen to be 0.5 m and 2 m below the ground surface, respectively. Five other different values of groundwater levels were studied separately. The effect of loss in the slurry pressure due to the existence of a weak soil layer at a certain depth of the trench was simulated by reducing the slurry pressure (SP) at some depths. The pile group was located at a distance (x) from the trench. This distance was considered to be 3.5 m in case of studying the other parameters.

3.2 Results from the Parametric Study

The effect of each parameter on piles deflection and bending moment is shown in Figs. 11, 12, 13, 14, 15 and 16, such an effect was generally presented for the first pile within the group. The effect of change of panel length on the pile behavior was very high but the effect of change in panel thickness shows was very low as shown in Fig. 11.

The location of the pile within the group and its distance from the trench controls the values and shape of the pile deflection and bending moment as shown in Figs. 12 and 13. The piles nearest to the trench within the group (i.e. piles 1 and 2) their tip deflect more than their top, because they move with the soil while the rear piles (i.e. piles 3 and 4) deflect from their top higher because they are affected by the drag force

Table 3. Piles properties used in the parametric study

Pile diameter	Pile				Interface			
	Area (m^2)	Inertia (m^4)	Perimeter (m)	E (kN/m^2)	k_s	k_n	ϕ_s	ϕ_n
					(kN/m^3)		(degree)	
0.2	0.0314	0.00015	0.6283	2.5×10^7	Equation 3		$2/3\phi$	
0.4	0.125	0.0025	1.25					
0.6	0.283	0.0120	1.88					
0.8	0.502	0.0402	2.51					
1.0	0.785	0.0981	3.14					
1.2	1.131	0.203	3.77					

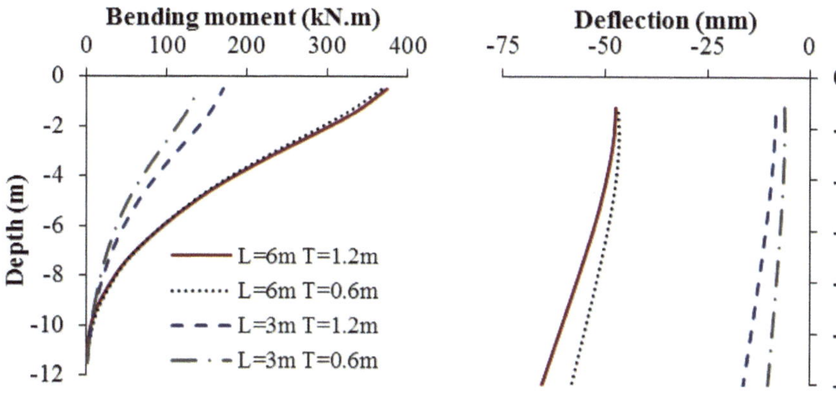

Fig. 11. Panels with different lengths and thickness

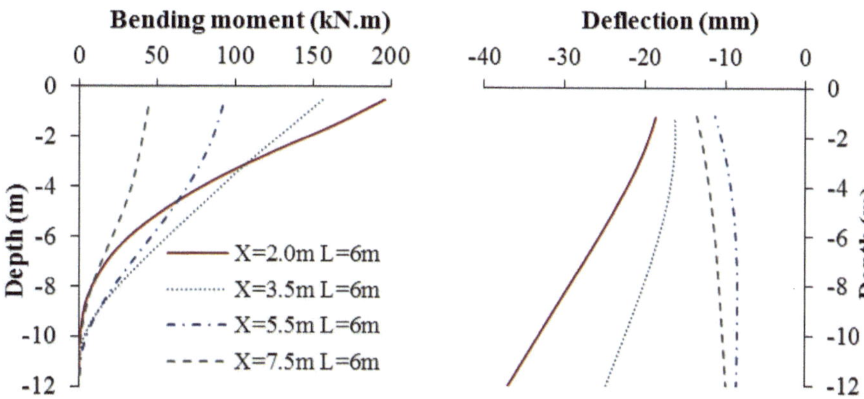

Fig. 12. Piles at different distances (x) from the trench

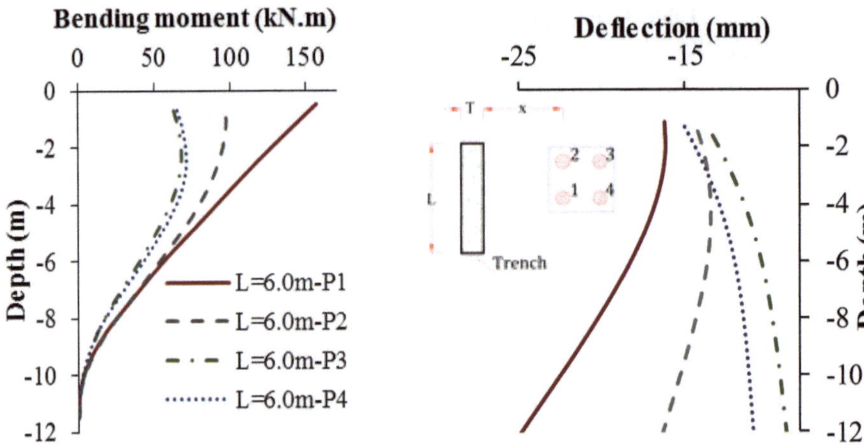

Fig. 13. Piles at different positions within the group

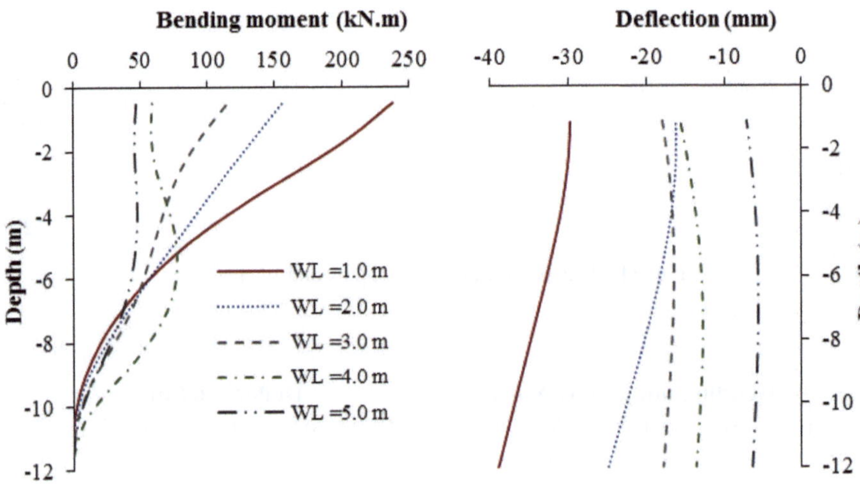

Fig. 14. Different values of ground water level

of the front pile. Generally, the deflection and bending moment values decrease with the distance from the trench.

The effect of different groundwater levels on the first pile within a pile group is presented in Fig. 14. The deeper the groundwater level the lower the deflection and bending moment. A noticeable difference was found when the groundwater table was 1.0 m below the ground surface. The effect of the slurry pressure reduction was higher than the effect of change in the groundwater level as shown in Fig. 15. The pile was greatly affected when the slurry pressure was reduced at depths between 11.5 m and 12.5 m (i.e. the location of the pile tip).

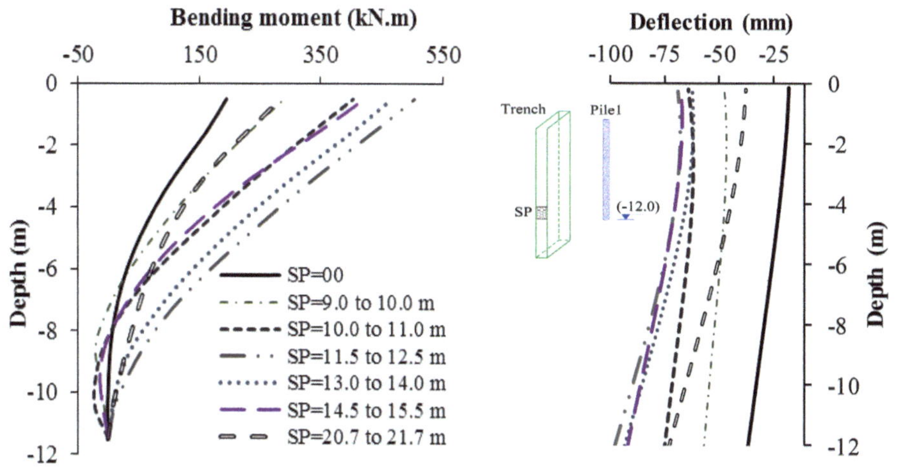

Fig. 15. Piles at different positions within the group

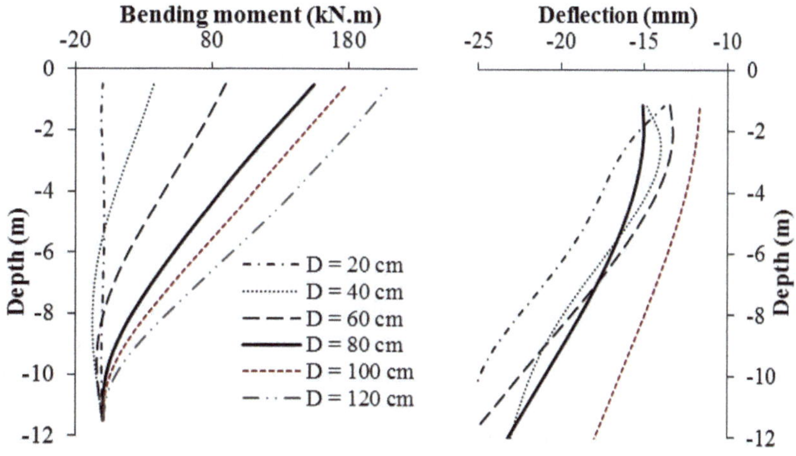

Fig. 16. Piles with different diameters

The large diameter pile is associated with low values of deflection but high values of bending moment as shown in Fig. 16. The large diameter pile provides a high stiffness and hence a high resistance to the soil movement. The ratio of change of bending moment values associated with the decrease in deflection is almost constant but the difference in deflection values was not constant. The deflection at pile diameters of 40, 60 and 80 cm was almost the same but it was slightly higher at a pile diameter of 20 cm and relatively low for a pile diameter of 100 cm.

The effect of panel construction stages on the first and third piles within the group is shown in Fig. 17. The values of the pile deflection and bending moment due to trenching the panel in two stages were lower than that due to trenching the panel in one stage.

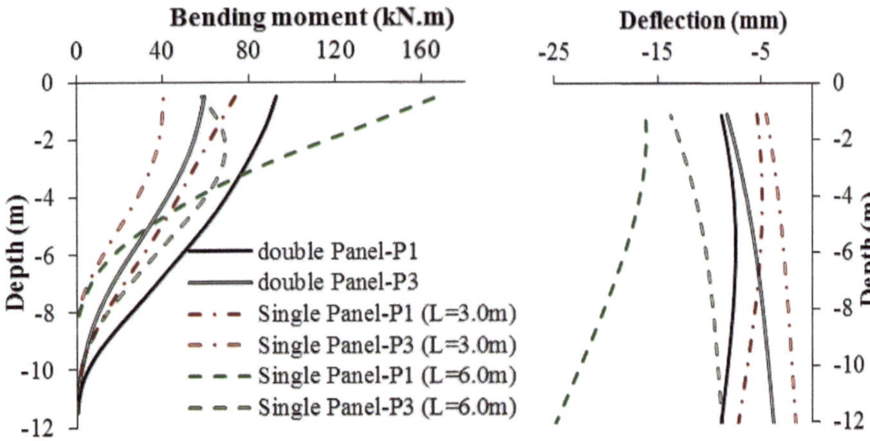

Fig. 17. Panel construction stages

4 Conclusions

The three-dimensional numerical simulation of the diaphragm wall trenching process was verified using two different case histories. The settlement results from the numerical analysis and field data were in a good agreement. However, some differences between both results were found due to the random nature of the field data results, while the numerical analysis depends on mathematical relations and provides systematic results.

The pile deflection and bending moment due to trenching were presented through a parametric study. The different values of panel thickness and pile diameter did not show a noticeable difference in the pile deflection, but the different values of pile diameter causes a noticeable change in pile bending moment. On the other hand, the change in panel length is associated with a great change in pile deflection and bending moment.

The change in values of groundwater level and pile group location causes a noticeable effect on the pile behavior, while the reduction of slurry pressure at some levels causes a great effect on the pile deflection and bending moment.

The piles within the group behave differently according to their position within the pile group. Generally, the piles nearest to the trench show higher deflection than those far from the trench and they are affected by the soil movement while the rear piles are affected by the drag force from the front piles.

The effect of trenching on piles could be reduced by conducting the panel in two stages, using shorter panel lengths and controlling the slurry pressure. It also could be reduced by lowering the groundwater table in the area before trenching.

References

Abdel-Rahman, A.H., El-Sayed, S.M.: Foundation subsidence due to trenching of diaphragm walls and deep braced excavations in alluvium soils. In: Proceedings of the 17th International Geotechnical Engineering Conference, vol. 3, pp. 1935–1938. Alexandria, Egypt (2009):

Byrne, P.M., Park, S.S., Beaty, M.: Seismic liquefaction: centrifuge and numerical modeling. In: FLAC and Numerical Modeling in Geomechanics: Proceedings of the 3rd International FLAC Symposium, Sudbury, Ontario, Canada, pp. 321–331 (2003)

Choy, C.K., Standing, J.R., Mair, R.J.: Stability of a loaded pile adjacent to a slurry-supported trench. Géotechnique **57**(10), 807–819 (2007)

Comodromos, E.M., Papadpoulou, M.C., Konstantinidis, G.K.: Effects from diaphragm wall installation to surrounding soil and adjacent buildings. Comput. Geotech. **53**, 106–121 (2013)

Cowland, J.W.; Thorley, C.B.B.: Ground and building settlement associated with adjacent slurry trench excavation. In: Proceedings of the Third International Conference on Ground Movements and Structures, pp. 723–738. Cardiff, England (1985)

Davies, R.V., Henkel, D.: Geotechnical problems associated with the construction of Charter station, Hong Kong. Arup J. **17**(1), 4–10 (1982)

DiBiagio, E., and Myrvoll, F.: Full scale field tests of a slurry trench excavation in soft clay. In: Proceedings of the 5th European Conference on Soil Mechanics and Foundation Engineering, Madrid, vol. 1, pp. 473–483 (1972)

Gourvenec, S.M., Powrie, W.: Three-dimensional finite-element analysis of diaphragm wall installation. Géotechnique **49**(6), 801–823 (1999)

Grandas-Tavera, C.E., Triantafyllidis, T.: Simulation of a corner slurry trench failure in clay. Comput. Geotech. **45**, 107–117 (2012)

Hamza, M.M., Atta, A., Roussin, A.: Ground movements due to the construction of cut-and-cover structures and slurry shield tunnel of the cairo metro. Tunn. Undergr. Space Technol. **14**(3), 281–289 (1999)

Itasca: Fast Lagrangian Analysis of Continua in 3 Dimensions (FLAC 3D) version 5 User's Guide. 5th edn. Itasca Consulting Group Inc., Minneapolis (2012)

Karlsrud, K.: Performance and design of slurry walls in soft clay. Nor. Geotech. Inst. Oslo **149**, 1–9 (1983)

Korff, M.: Response of piled buildings to the construction of deep excavations. Ph.D. Dissertation. University of Cambridge, UK (2013). http://www.dspace.cam.ac.uk/handle/1810/244715

L'Amante, D., Flora, A., Russo, G., Viggiani, C.: Displacements induced by the installation of diaphragm panels. Acta Geotech. **7**, 203–218 (2012)

Mohamed, A.: Effect of groundwater table rising and slurry reduction during diaphragm wall trenching on stability of adjacent piles. In: IOP Conference Series: Earth and Environmental Science, Conference, vol. 1, p. 26 (2015). doi:10.1088/1755-1315/26/1/012012

Ng, C.W.W., Yan, R.W.M.: Stress transfer and deformation around a diaphragm wall. J. Geotech. Geoenviron. Eng. **124**(7), 638–648 (1998)

Ng, C.W.W., Rigby, D., Lei, G.H., Ng, S.W.L.: Observed performance of a short diaphragm wall panel. Géotechnique **49**(5), 681–694 (1999)

Poh, T.Y., Goh, A.T.C., Wong, I.H.: Ground movements associated with wall construction: case histories. J. Geotech. Geoenviron. Eng. **127**(12), 1061–1069 (2001)

Stround, M.A., Sweeney, D.J.: A review of Diaphragm wall. Discussion Appendix. Institution of Civil Engineers (1977)

Tsai, J.-S., Jou, L.-D., Hsieh, H.-S.: A full-scale stability experiment on a diaphragm wall trench. Can. Geotech. J. **37**(2), 379–392 (2000)

Treatment of a Landslide by Using Piles System, Case Study of the East-West Highway of Algeria

Sidi Mohammed El-Amine Bourdim[1(✉)], Lokmane El-Hakim Chekroun[2], Abdelkader Benanane[1], and Abdelillah Bourdim[3]

[1] LMPC Laboratory, Department of Civil Engineering and Architecture, University of Mostaganem, Mostaganem, Algeria
sidimohammed.bourdim@univ-mosta.dz
[2] EOLE Laboratory, Department of Civil Engineering, University of Tlemcen, Tlemcen, Algeria
[3] National Agency of Highway, Office of Tlemcen, Tlemcen, Algeria

Abstract. The use of piles to slope stability has grown in recent years through the good reported performance/time offered by this technique. In this context, we present the case of a landslide that occurred on 2nd March, 2014 on the highway East-West Algeria, a section of Tlemcen, using anti-sliding piles as the treatment solution.

Our study case (landslide) occurred near the village of Ouled Mimoun (Tlemcen), specifically between Pk52+040 ~ Pk52+220. This disorder appeared in the form of longitudinal cracks on the pavement about 50 m of length. The left side of the roadway collapsed following the slip of the downstream slope.

Surveys and Inclinometers were installed to follow the deformation of the slope in time and they showed signs of instability with a sliding depth of about 9 m near the motorway platform.

The probable causes of this instability are the removal of the bottom abutment for the upstream slope of the highway and the establishment of an earth deposit that was overloaded the slope and disrupted the flow of waters to the downstream of Pk52+000.

The study of stabilization of this slide is based on the installation of two lines of anti-sliding piles and the introduction of geosynthetic drainage system (large draining spurs made in the direction of the slope). Our study of stability analysis was carried out under static and dynamic loads and highlights that this solution is advantageous and efficient. We note that this solution was chosen by the company (Group CITIC-CRCC) chosen to repair this section of highway.

Keywords: Landslide · Stabilization · Anti-slidings piles · Dynamic study

T. Abdoun and S. Elfass (eds.), *Soil Dynamics and Soil-Structure Interaction for Resilient Infrastructure*, Sustainable Civil Infrastructures, DOI 10.1007/978-3-319-63543-9_2

1 Introduction

Il existe des principaux facteurs qui contrôlent le type et le taux de mouvements de masse qui pourrait se produire.

The origins of the loss of stability, both natural and artificial, are very diverse. The setting in motion of natural slopes (slow or sudden) may cause significant damage to structures, with a significant economic impact, and sometimes injury or death to humans. The study of a slope comprises, besides the recognition of the site and the selection of the mechanical characteristics of the soil, a stability calculation to determine firstly the failure curve along which the risk of slip is highest, the surface along which the factor of safety is the lowest. Landslides are diverse, both in their nature and size. Slope stability problems usually occur in the realization of roads, dikes of dams and natural slopes. The landslide passes through several chronological stages of activity. There are main factors that control the type and rate of mass movements that might occur.

The method of stabilization of the slopes by piles, the so-called anti-sliding piles, have attracted the interest of several researchers. Many studies have been conducted on the stability of slopes after introducing stabilizing piles. There are two calculation methods: limit equilibrium methods and numerical methods [1]. The kinematic approach dealing the slope with more rows of piles called "passive piles" has the effect of improving its stability. Recall that the major problem in the design of slope-pile system is determining the proper location of the piles.

We present the case of a landslide that occurred on a highway section of the East-West highway of Algeria. The landslide was treated using anti-sliding piles to reinforce the embankment seat of the body pavement. Damage and cracks were observed for the first time on the Tlemcen-Algiers route at PK 52 in November 2013. Thereafter, the pavement, became very degraded and then useless in March, 2014.

2 Presentation of the Landslide

In this article we discuss the landslide which occurred on March 2, 2014 on the section of the East-West highway of Algeria located near the town of Tlemcen (West of Algeria). Indeed, the platform at Pk52+040 ~ Pk52+093 suffered significant deformations, where the back wing moved vertically over 2 m and 3 to 4 m horizontally. Thus the left side of the road was completely stopped for traffic as it shows the Photo 1.

In PK52+093 ~ 200, the outside was completely distorted; it was subject to sliding. According to some reports, the landslide developed quickly with dense cracks in the body of the slide near the river, evident distortions in the form of shear.

Data analysis based on the inclinometer readings showed that the sliding surface was at 9 m, 7 m and 3 m as shown in Table 1 [2].

The displacements at the foot of the slope on the right side of the platform showed no sign of significant deformation during the three months following the event date.

Photo 1. Deformation of the pavement body

Table 1. Inclinometer readings at level of the deformation zone.

N° of drilling	Date of first measurement	Date of the end of measurement	Sliding depth (m)	Cumulative displacement (mm)
S01	13-12-2013	29-12-2014	9	26.99
S02	18-12-2013	30-12-2013	7	–
S03	25-12-2103	02-03-2014	3	58.81

3 Landslide Causes

The vertical displacement of the sliding object platform is about 2 m, a fill section. Observing the collected core samples showed a very variable lithology, containing a large pebble content in holes drilled. According to the order of destruction of inclinometer surveys, the borehole located next to the bank of the river was the first hole damaged, and that is the hole near the destroyed platform. The slide then develops due to traction from the bottom upwards. In addition, there is no deformation in the hole that is on the right side of the roadway. Waters of the wadi on the front edge of the slide quickly caused the appearance scour.

A day before the drama, i.e. 03/01/2014, torrential rains lashed the region and caused seepage water and a rise in the level of groundwater in the body of the bank, providing a lubricating action that has accelerated the phenomenon of the sliding mechanism.

4 Geological and Geotechnical Profile Site

The geotechnical profile of the site as determined by reconnaissance, six boreholes to a total depth of 127.4 m, shows that the area has a sloping profile. The soil layers consist of backfill of average density, followed by mainly disturbed brownish to yellowish silty clay in the trough left of the plot. This plastic soil is homogeneous interspersed with little pebbles and sand. Shale beds have completely altered to very plastic green yellow clay structure with sandwiched sand. The structure is of variable depth of 13 to 24.5 m in different boreholes with a tendency to flow in the direction of the slope (Fig. 1).

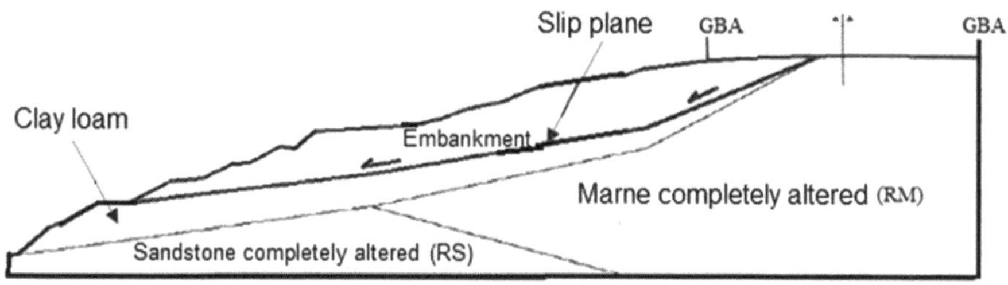

Fig. 1. Geotechnical profile site

Table 2 summarizes the values of the geotechnical properties of soil layers provided by the company of the laboratory conducting the project (CITIC-CRCC).

Table 2. Geotechnical parameters of soil layers [2].

Sol	Volumic weight γ (kN/m^3)	Drained cohesion C (kPa)	Friction angle (°)	Natural water content *w* (%)	Liquidity limit *wl* (%)	Plasticity limit *wp* (%)	Void ratio *e*
Backfill	19.2–19.9	15	27	21.6–23.4	–	–	–
Silty clay	18.2–20.8	14.7	22	20.4–28.7	53.1–58.2	21.9–27.5	0.59–0.92
Marne completely altered (RM)	18.3–20.5	14.7	22	13.9–24.8	26.0–57.5	14.5–23.5	0.53–0.80

For the underlying layer of sandstone that is completely altered (RS), no value has been provided by the laboratory, but it was mentioned that it is a fine sandstone of bad rock characteristics and belongs to the category relatively soft rock.

5 Treatment of Landslide

A static calculation of the stability of the slope using code Plaxis 2D [3] showed that the slip circle passes through the artificial slope (see Figs. 2, 3, 4 and 5).

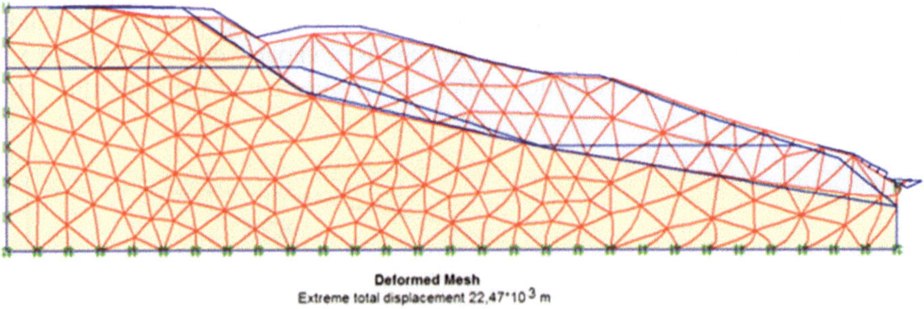

Fig. 2. Mesh of deformation due to self weight of the slope

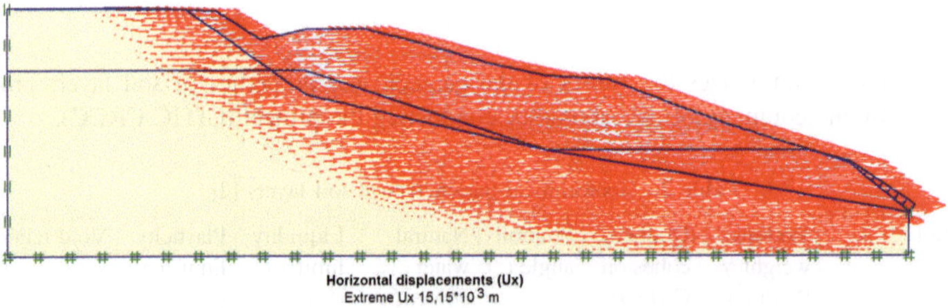

Fig. 3. Fields of horizontal displacements

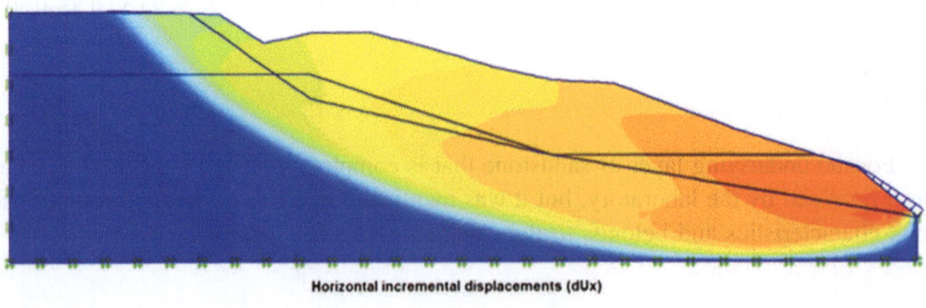

Fig. 4. Critical slip surface

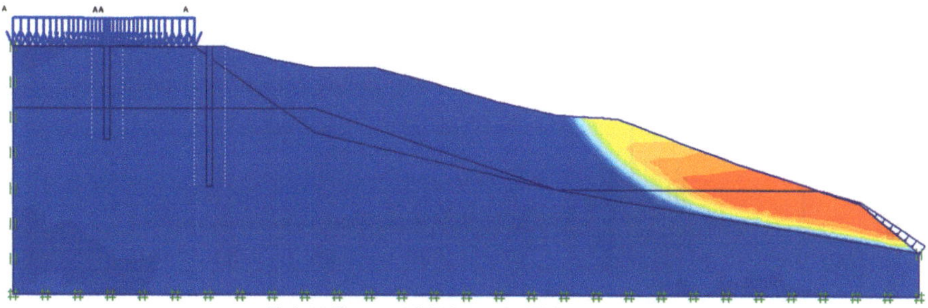

Fig. 5. Failure mechanism and sliding surface after introduction of the piles

The solution chosen for the treatment involves the installation of 11 emergency anti-slip piles in a single row at the edge of the slip surface (middle of the road) to ensure the safety of users of the highway. This is followed by the installation of a row of 60 piles on the left side of Pk52+040 ~ 220 PK52 with a beam connecting the piles together at the heads.

The piles have a diameter of 1.2 m, while the depth is 15 m for the first line of the 11 stakes and 19 to 23 m for the second line of the 60 piles. They are spaced 3 m apart. The design strength of the concrete used for the piles is 35 MPa while the Young's modulus is E = 35982 MPa. The specific gravity used for the concrete equals 25 kN/m^3. Soil-concrete contact interface elements were used with reduced properties (2/3 compared to that of the soil).

The Fig. 5 shows an analysis of the stability of the slide after the introduction of two rows of anti-sliding piles. Operating overload vehicles on the road are taken equal to 10 kN/m^2 are arranged 1.50 m from the edge of the roadway. It is clear that the bank is more stable compared to the previous configuration with increased safety factor. The roadway is now in a safe condition.

6 Dynamic Analysis

The slip treatment proposal was the subject of plane strain static and dynamic modeling using the PLAXIS 2D Version 8.2 software [3, 4]. It was subjected to a combination of load (dead load, operating load and seismic load).

The seismic loading applied is that of the 2003 Boumerdes earthquake (May 21st), characterized by a magnitude of 6.8 and a PGA (Peak Ground Acceleration) of 0.34 g [5]. The input signal was the E-W component recorded at Kedarra Station (Fig. 6).

Note that the numerical model used was calibrated to previous work [6]. Several suggestions that different investigators made relative to Plaxis parameters such as the choice of the digital model borders [7], system damping [8], coefficients of Newmark [8], refinement degree of the mesh [9] and the number of dynamic sub steps [8] were adopted.

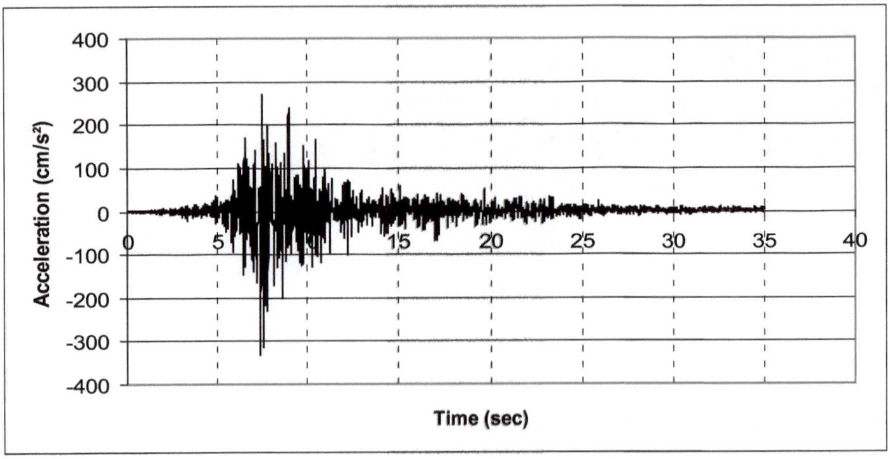

Fig. 6. The seismic loading applied

Although various simulations were performed, we present only the deformed mesh of the configuration at the end of seismic loading (t = 35 s). This deformation is shown in Fig. 7.

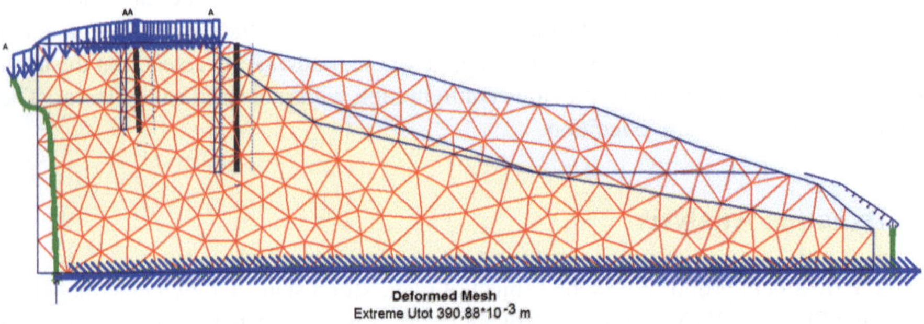

Fig. 7. Deformed mesh of the slide with piles

Figure 8 represents the distribution of horizontal displacements in the slope body. Figure 9 illustrates the relative shear stresses at the end of the seismic loading for the case of reinforcement of the slope with piles.

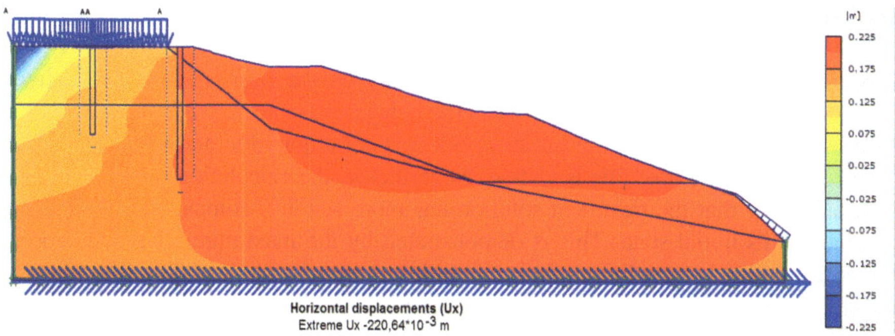

Fig. 8. Field of horizontal displacement with maximum value of 22 cm

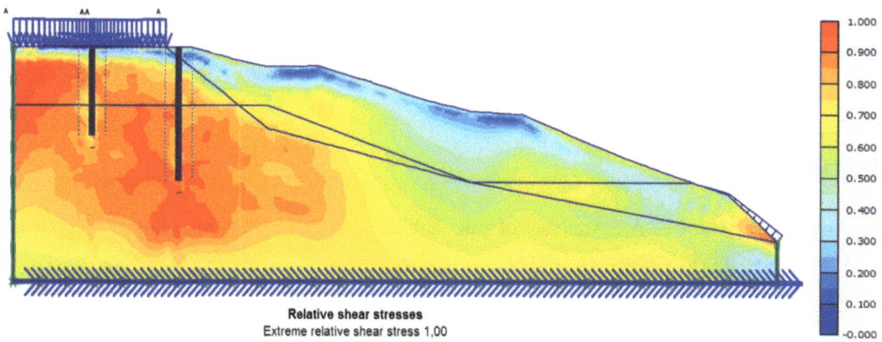

Fig. 9. Relative shear stresses in the case of the slide with piles

7 Discussion

While the time calculation is not given here, it was found that the maximum values of displacements and stresses correspond to the peak acceleration (PGA = 0.34 g arriving at t = 8 s). Amplification begins with the arrival of shear waves.

For deformation, the recorded maximum displacement is 22 cm. This value is negligible vis-a-vis the large scale of the slope. For shear stresses, Fig. 9 shows that the values are relatively remote from the failure except in some areas around the pile that is to say, soil-piles interface they are very close to the plastic limit. This is due to reduced interface parameters (soil cohesion and friction piles) that are taken in the calculations, which are very low [10].

8 Conclusion

Vertical piles were used to stabilize the roadway on the right side of the highway section of PK 52 in the boundary between the wilaya of Tlemcen and Sidi Bel Abbes in Algeria. The company carrying out the work was not required to adopt our analysis to establish the treatment of this landslide. Nevertheless the calculations, both static and dynamic, show that the treatment solution has more stability compared to the case of the slope in its initial state. This is demonstrated by the maximum value of recorded movement which is very low and the safety factor found to be equal to 1.86 after the introduction of the piles. If we add to the benefits of this solution compared to other conventional solutions, the quantities of materials and construction time saved, the pile solution is very advantageous.

Based on studies of the stabilization of PK 52 slopes with dynamic analysis using an accelerogram of strong seismic movement, Boumerdes 2003, we conclude that slope stabilization with piles may permanently resolve the problem of landslide. This technique is primarily applicable to slopes of clay soils, sometimes soft or sensitive.

References

1. Ashour, M., Hamed, A.: Analysis of pile stabilized slopes based on soil–pile interaction. Comput. Geotech. **39**, 85–97 (2012)
2. CITIC-CRCC: Plan de traitement du glissement de PK52+040 ~ 220, Indice G. Projet Autoroute Est-Ouest, Algérie (2014)
3. Brinkgereve, R., Vermeer, P.: PLAXIS Version 8, Scientific manual. DELFT University of Technology & PLAXIS BV, Pays-Bas (2003)
4. Brinkgereve, R., Vermeer, P.: PLAXIS Version 8, Dynamics manual. DELFT University of Technology & PLAXIS BV, Pays-Bas (2003)
5. Laouami, N., Slimani, A., Bouhadad, Y., Chatelain, J.L., Nour, A.: Evidence for fault-related directionality and localized site effects from strong motion recordings of the 2003 Boumerdes (Algeria) earthquake: Consequences ondamage distribution and the Algerian seismic code. Soil Dyn. Earthq. Eng. **26**, 991–1003 (2006)
6. Bourdim, S.M.A., Djedid, A., Boumechra, N.: Treatement of a landsilde on the section of east-west highway in Algeria. In: 6th Engineering and Technology Symposium, April 25–26, 2013. Çankaya University, Ankara (2013). ISBN: 978-975-6734-15-5
7. Yun-Suk, C.: Etude numérique de l'interaction sol-pieu-structure sous chargement sismique. Thèse de Doctorat, Université des Sciences et Technologies de Lille, France (2000)
8. Bourdim, S.M.A., Boumechra, N., Djedid, A.: Numerical model calibration, case of dynamic behavior of a soil-retaining wall system. J. Mater. Environ. Sci. **7**(3), 1048–1055 (2016)
9. Kuhlemeyer, R., Lysmer, J.: Soil Mech. Found. Div. **99**, 421–427 (1973)
10. Chekroun, L.H., Boumechra, N., Djedid, A.: Behavior of a pile group (3 $\times$ 3) subjected to lateral loading. J. Mater. Environ. Sci. **6**(11), 3319–3328 (2015)

Centrifuge Modeling of Mine Tailings and Waste Rock Co-disposal, Consolidation and Dynamic Loading

Nonika Antonaki[1(✉)], Tarek Abdoun[2], and Inthuorn Sasanakul[3]

[1] Geotechnical Engineer, WSP-Parsons Brinckerhoff, New York, NY 10119, USA
antonakin@pbworld.com, nonika.a@live.com
[2] CEE, RPI, Troy, NY 12180, USA
[3] CEE, University of South Carolina, Columbia, SC 29208, USA

Abstract. Tailings from a planned copper-gold mining project located at the earthquake-prone Andean region of South America were obtained from the metallurgical pilot plant in order to perform centrifuge tests at the Center for Earthquake Engineering Simulation (RPI). Consolidation and shaking table tests were conducted to evaluate the properties and liquefaction potential of mildly sloped consolidated mine tailings. A gentle slope at the surface of a tailings impoundment can significantly add to the stored volume. In the field, the tailings are thickened and hydraulically deposited into the containment structure in layers. In the centrifuge, tailings were prepared in layers and consolidated, thus allowing instrumentation of each layer before consolidation of the complete impoundment. Due to long consolidation time, large settlement and clear signs of liquefaction after a few cycles of dynamic loading, the need for improvement arose. One of the alternative management methods that can improve physical stability and geochemical properties is co-disposal of mine tailings and waste rock. In this study, co-mixing of the materials at a specified ratio of dry mass (waste rock to tailings) prior to disposal was examined. The behavior was compared to that of tailings alone with respect to consolidation rate, settlement accumulation, slope stability and response to dynamic loading.

1 Introduction

Conventional mining activity produces large quantities of both tailings with high water content and dry waste rock. Tailings are commonly discharged as a slurry with slow consolidation properties and low shear strength, often causing failures in tailings impoundments due to physical instability. In contrast, the waste rock is characterized by high shear strength and is commonly disposed of in large dumps. The unsaturated conditions allow weathering of the waste rock, which may cause long-term acid drainage and metal leaching. Blending the two materials can produce self-sealing deposits with high shear strength, low compressibility and density higher than either material on its own, thereby reducing the total volume and surface area requirements for impoundment design and final reclamation (Bussiere 2007; Wilson et al. 2009).

T. Abdoun and S. Elfass (eds.), *Soil Dynamics and Soil-Structure Interaction for Resilient Infrastructure*, Sustainable Civil Infrastructures, DOI 10.1007/978-3-319-63543-9_3

To date, most experimental research on mine tailings has consisted of small scale laboratory tests, such as simple or direct shear tests, triaxial tests, etc. (Aubertin et al. 1996; Dimitrova et al. 2011; Garga et al. 1984; Highter and Tobin 1980; Highter and Vallee 1980; Proskin et al. 2010; Qiu et al. 2001; Sanin et al. 2012; Wijewickreme et al. 2005; etc.). Laboratory investigations have also been conducted on mixtures by a few researchers (Leduc et al. 2003; Wickland et al. 2005; Wickland et al. 2006; Wickland et al. 2010). In Australia, the co-disposal of coarse and fine coal wastes by combined pumping was pioneered in 1990 by Jeebropilly colliery in southeastern Queensland (Morris et al. 1997). Very few attempts of physical modeling tests using a centrifuge have been made with mine tailings (Stone et al. 1994).

In this study, fine-grained tailings from a planned copper-gold mine in an earthquake-prone zone of South America were used in centrifuge tests. The main objectives were to model the deposition of tailings in layers on the centrifuge, examine the consolidation behavior in terms of time, pore pressure dissipation rate and settlement and, finally, to evaluate the liquefaction potential and slope stability under dynamic loading. When the material is able to sustain a sloped surface the stored volume within a specific area increases.

Subsequently, the same testing scenario was repeated using a mixture at a selected ratio of waste rock to tailings by dry mass for comparison purposes. Geotechnical centrifuge testing can offer several advantages. Consolidation is accelerated and simulating prototype stress conditions in a small scale model is enabled (Taylor 1995). A new technique for monitoring settlement of the layers during consolidation was also developed and has been presented by Antonaki et al. (2014).

2 Test Description

A series of tests were carried out at the Center for Earthquake Engineering Simulation at Rensselaer Polytechnic Institute (Troy, NY). The centrifuge is a 150 g-ton machine with a nominal radius of 2.7 m. These tests were conducted at 80 g centrifugal acceleration to model a 20 to 25 m high deposit. In an attempt to simulate field conditions the models were constructed in four layers. Drainage was only allowed at the surface to model the in situ conditions.

The tailings were shipped from the site thickened to 59% water content. The material had a specific gravity of 2.73, about 60% passing through the #200 mesh and was classified as low-plastic (CL-ML). For the centrifuge tests, initial water content was re-adjusted to the field pumping water content of 59%. Uniform crushed stone (1/8") was purchased and then sieved using a 0.203" sieve to remove particles larger than 5 mm or about 37 cm in prototype scale. Measurements and calculations resulted in 2.3:1 (rock waste to mine tailings by dry mass) as the ratio leading to fine material just filling the voids of the waste rock skeleton at the consolidated state, which is considered to be the optimal scenario when designing a mixture (Wickland et al. 2006). The 2:1 mixture presented herein was selected as a starting point for a parametric study.

The deformable laminar container with internal dimensions of 71 cm × 36 cm 37 cm that was used is depicted in Fig. 1. This type of container consists of stacked thin rectangular "rings" which can slide with respect to each other on bearings, thus allowing the model to deform laterally as opposed to imposing unrealistic fixed boundary conditions during dynamic loading. Removable rigid side walls ensured fixed boundaries during the deposition and consolidation phases of the tests. Figure 2 (a) depicts the container on the centrifuge while its rigid ends are still attached. A membrane fitted to the inside of the container was fabricated to safely contain the material, since the interface between rings cannot prevent seepage of water.

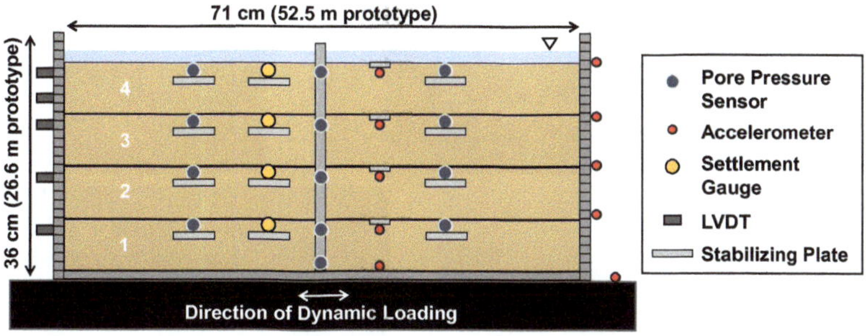

Fig. 1. Test set-up, layer numbering and sensors used.

The final instrumentation setup is shown in Fig. 1. Settlement gauges, pore pressure sensors, accelerometers and LVDTs were used to obtain data corresponding to each layer. Pore pressures and settlements were monitored to evaluate the consolidation progress with time and depth. The consolidation phase of the tests could conclude once these parameters became practically constant with time, especially relying on pore pressure transducers fixed on the container wall, as depicted in Fig. 2 (b). Measuring tape was glued on the fabricated membrane in order to observe settlement throughout the tests (Fig. 2 (b)). Accelerometers were embedded below the surface of each layer, as well as glued on the rings of the container at the level of each layer surface after consolidation, as estimated from previous tests. LVDTs were used to monitor lateral deformation of the container during dynamic loading. All sensors placed inside the soil model needed to be glued to small plastic or aluminum plates, in order to maintain their position and direction throughout the tests (Fig. 2 (c), (d)). Accelerometer cables were additionally wrapped in aluminum foil while the rest of the sensor plates went through vertical strings that prevented lateral movement. During the tests, videos were recorded via centrifuge onboard cameras.

All tests consisted of three phases: brief consolidation of each deposited layer before deposition of the next, consolidation of the complete model and dynamic loading of the sloped model. The first phase was conducted for 30 h (prototype time) at a lower g-level (20 g). This procedure allowed the material to gain enough strength for sensors to be placed at the surface. At the same time the short-term consolidation of tailings as the impoundment is gradually filled was modelled. Consolidation of the impoundment at 80 g centrifugal acceleration followed, with duration depending on the material being tested. After consolidation was completed water was drained from the surface of the model and a mild slope ($\approx$ 4%) was formed. Thickened tailings can be

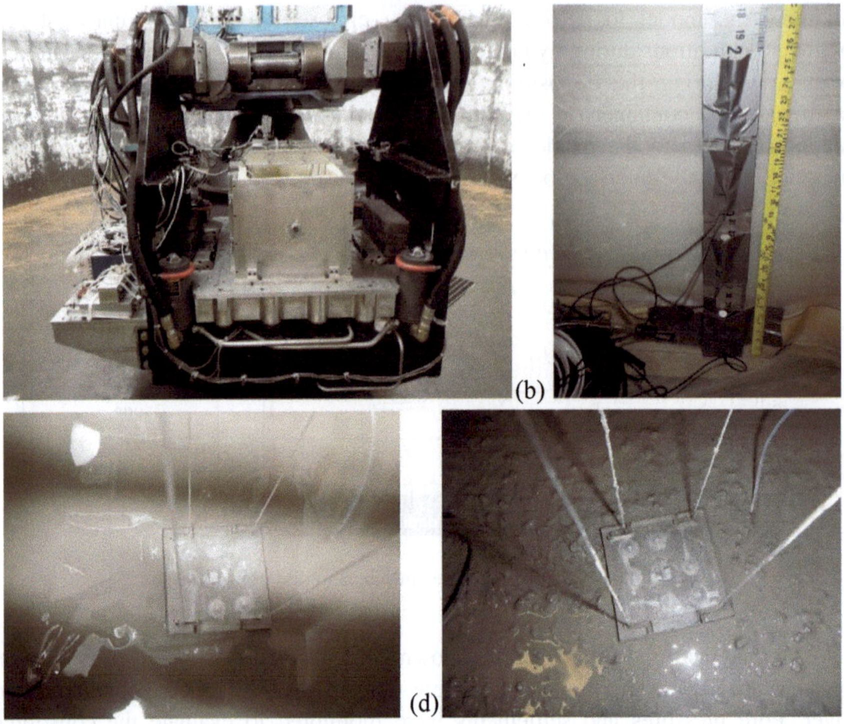

(a) (b) (c) (d)

Fig. 2. (a) Laminar container bolted on shaker attached to the centrifuge basket, (b) pore pressure sensors taped on ruler attached to the wall of the container and measuring tape used to monitor settlement, (c) sensor glued on plastic plate floating on surface of mine tailings and (d) same sensor on surface of 2:1 mixture.

deposited to form a gentle beach slope (typically between 2% and 6%). The model was spun back to 80 g and a harmonic motion was applied at the base of the model by the shaking table (Fig. 3 (a), (b)). The motion consisted of 50 cycles, with linearly increasing amplitude that reached its maximum (approximately 0.10 g) after the first 10 cycles. Figure 3 (c) depicts the container on the shaker after the rigid sides were removed prior to dynamic loading. The LVDT set-up can also be observed in the same figure. Figure 3 (d) exhibits the deformed container at the end of a test. After completion of each test the model was carefully dissected in order to identify the final location of sensors and collect soil samples for water content measurements. Figure 3 (e) and (f) depict accelerometers during dissection after a test with just mine tailings and the 2:1 mixture respectively. The specific sensors successfully maintained their orientation, but that was not the case for all sensors placed in mine tailings. Sensor stability was more easily achieved in mixtures.

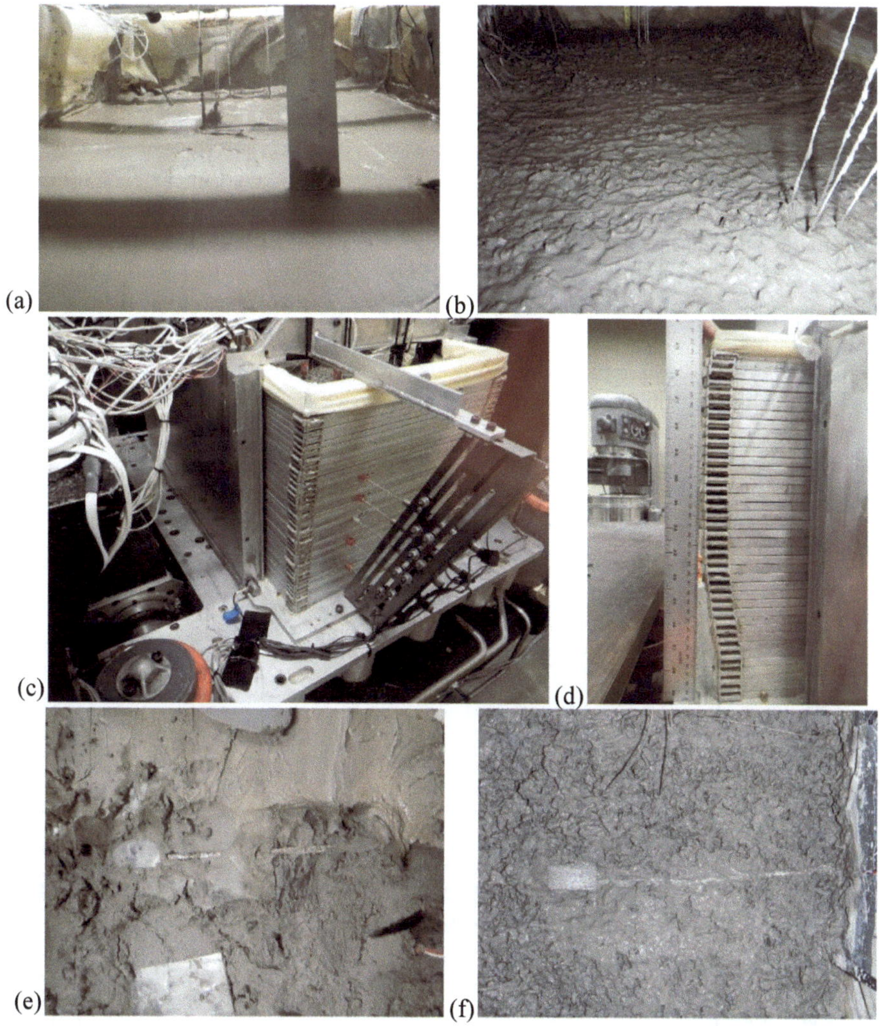

Fig. 3. (a) Laminar container after removal of rigid sides and LVDT set-up on sliding container rings, (b) example of deformed soil profile after completion of a test, (c) sloped surface of mine tailings before shaking, (d) sloped surface of 2:1 mixture before shaking, (e) accelerometer successfully stabilized in mine tailings until post testing dissection and (f) the same in 2:1 mixture.

3 Test Results

Data from a test performed with mine tailings and a test performed with a 2:1 mixture are compared in prototype scale.

Figure 4 presents pore pressure dissipation as measured by two selected fixed pore pressure sensors (shown in Fig. 2 (b)) during consolidation of the deposits at 80 g.

Excess pore pressure dissipation at similar depths for both materials is plotted in each graph. Every pore pressure curve was normalized with the final value to eliminate the effect of small differences in sensor depth between the two tests (≈ 4 m). The top graph shows pore pressure measured in the third of four layers for both materials (4 − 8 m from the water surface) and the bottom graph shows pore pressure measured in the first or bottom layer (15 – 19 m from the water surface), as numbered in Fig. 1. Curves are qualitatively similar for both tests, consisting of an initial steep part followed by a much flatter part that asymptotically approaches the hydrostatic value. However, mine tailings started with significantly higher build-up and took much longer to consolidate. The mine tailings were left to consolidate for three times the consolidation time of the mixture. In both cases a small part of what appears as pore pressure dissipation was evaporation. As indicated by the final slope of the curves, mine tailings were still slowly consolidating three and a half years after deposition.

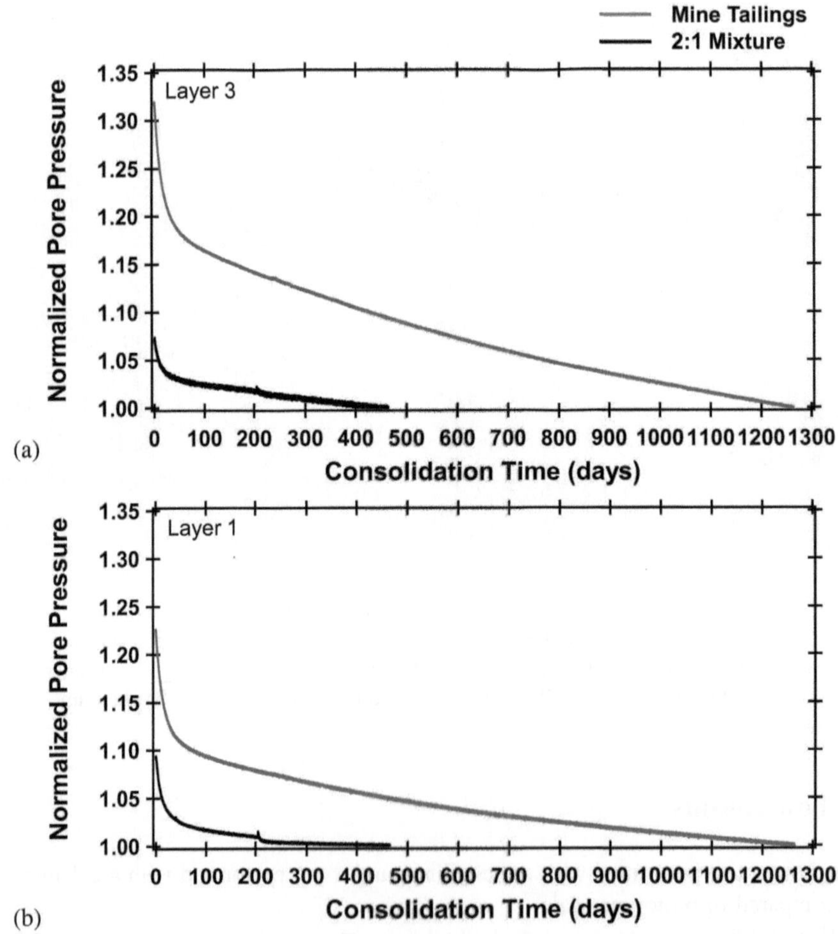

Fig. 4. Pore pressure dissipation (normalized with final value) measured in (a) third layer and (b) first layer during consolidation of mine tailings and 2:1 mixture.

For the purpose of dynamic response comparison, Figs. 5 and 6 depict selected acceleration and horizontal displacement data from the two tests. The top graph in Fig. 5 shows acceleration measured at the base of the container during dynamic loading of mine tailings. The maximum acceleration applied was approximately 0.096 g. The same excitation was applied to the mixture but the amplitude was 0.074 g due to variation of shaker response for different soil models. The middle and the bottom graphs in Fig. 5 correspond to acceleration measured right underneath the surface of the tailings and the mixture. In mine tailings, a visible drop in the acceleration data occurred even before the maximum base acceleration was reached and the acceleration became practically zero (less than 0.05 g) a few cycles later, indicating liquefaction of the slope. In contrast, the mixture surface acceleration kept increasing even after the maximum excitation amplitude was reached. A smaller and asymmetrical decrease was present in the acceleration data and indicated softening but not liquefaction. Accelerations measured at deeper layers are not included herein, but showed that liquefaction did not reach the bottom layer of the mine tailings deposit.

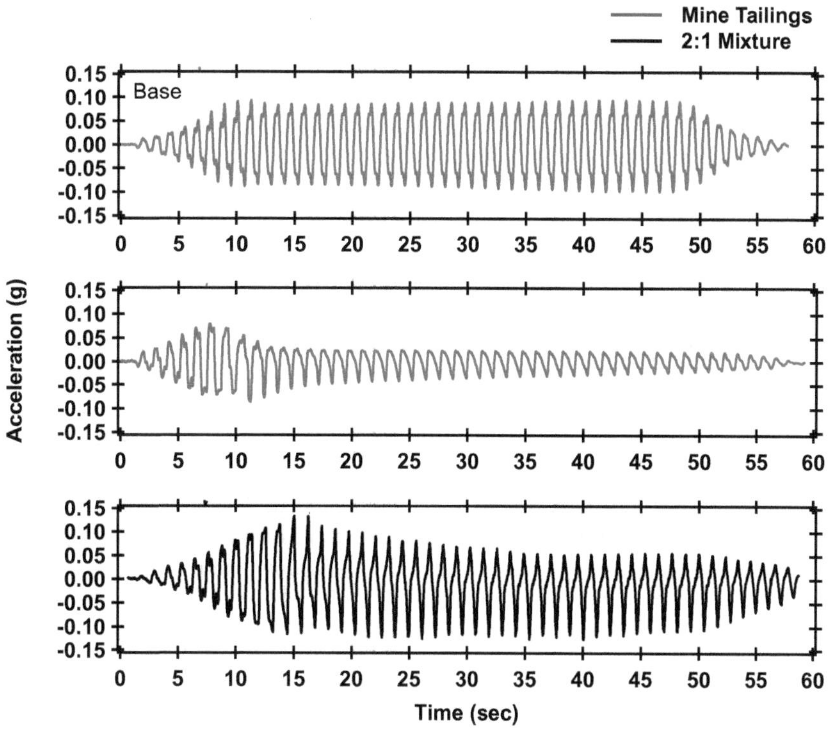

Fig. 5. Acceleration measured during dynamic loading: at base of mine tailings (top graph), at surface of mine tailings (middle graph) and at surface of 2:1 mixture (bottom graph).

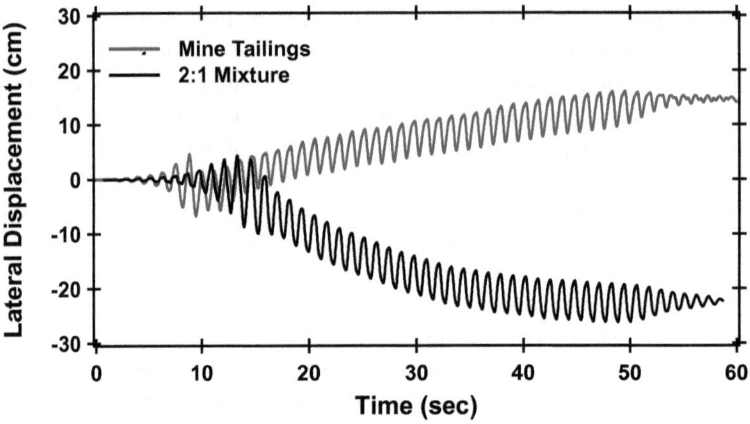

Fig. 6. Horizontal displacement measured at the surface level of mine tailings and 2:1 mixture during dynamic loading.

The behavior of the two deposits was similar with respect to lateral displacement of the soil profile. Displacement of the top of the deposits measured by LVDTs was not very large (Fig. 6). It appears that the residual displacement of the mixture was slightly higher, but the top LVDT was right at the level of the surface in that test and 1.5 m below the surface in the mine tailings test. The tendency has been for deformation to increase closer to the surface, so the residual displacements at the top can be considered comparable for the two tests. The mine tailings profile deformed in the direction towards the toe of the slope as expected, whereas the mixture profile deformed in the opposite direction. Deformations are small in both cases, so the direction is considered coincidental and not of great significance.

A summary of observed settlements is included in Table 1 for the two tests presented. Settlement was normalized with height and presented in percent. The last column of the table corresponds to the ratio between the settlements of the tailings (T) and the mixture (M) during each testing phase. The mine tailings settled 1.8 times more during the brief consolidation phase following deposition of each layer and 3.4 times more during consolidation of the entire deposit. Dynamic settlements were similar for the two materials. The negative settlement of the toe of the slope for the mixture indicates that some surficial material moved downslope. In the case of mine tailings the slope started at an inclination of 3.5% and ended up at 2.8% after the motion was applied, whereas in the case of the mixture the slope started at approximately 4.2% and ended up at 2.8%.

Table 1. Comparison between settlements normalized with model height of mine tailings and 2:1 mixture during each phase of the tests.

Testing phase	Layer	Settlement/Height (%)		Settlement ratio (T/M)
		Tailings	2:1 Mixture	
Consolidation at 20 g	1	15%	7%	2.2
	2	20%	13%	1.6
	3	22%	13%	1.6
	4	24%	12%	1.9
	Sum	20%	11%	1.8
Consolidation at 80 g	All	13%	4%	3.4
Shaking at 80 g	Head	4%	3%	1.2
	Toe	2%	−1%	−1.8

4 Conclusions

To summarize this study, mine tailings and a mixture of mine tailings and coarse waste rock at 2:1 ratio of waste rock to tailings by dry mass were deposited in layers, consolidated and dynamically loaded on the centrifuge. Co-disposal offered advantages regarding space requirements, consolidation time, compressibility and liquefaction potential. Slope stability was not significantly improved. Based on these results, more tests need to be conducted to examine mixture behavior and further improve the physical stability of the deposit. A centrifuge test with a 3:1 mixture was planned shortly after these conclusions were reached.

Acknowledgments. The authors appreciate the support of Golder Associates Ltd.

References

Antonaki, N., Sasanakul, I., Abdoun, T., Sanin, M.V., Puebla, H., Ubilla, J.: Centrifuge modeling of deposition and consolidation of fine-grained mine tailings. In: Proceedings of Geocongress 2014, Atlanta, Georgia (2014)

Aubertin, M., Bussiere, B., Chapuis, R.P.: Hydraulic conductivity of homogenized tailings from hard rock mines. Can. Geotech. J. **33**, 470–482 (1996)

Bussiere, B.: Colloquium 2004: hydrogeotechnical properties of hard rock tailings from metal mines and emerging geoenvironmental disposal approaches. Can. Geotech. J. **44**, 1019–1052 (2007)

Dimitrova, R.S., Yanful, E.K.: Undrained strength of deposited mine tailings beds: effect of water content, effective stress and time of consolidation. Geotech. Geol. Eng. **29**, 935–951 (2011)

Garga, V., McKay, L.D.: Cyclic triaxial strength of mine tailings. J. Geotech. Eng. **110**(8), 1091–1105 (1984)

Highter, W.H., Tobin, R.F.: Flow slides and the undrained brittleness index of some mine tailings. Eng. Geol. **16**, 71–82 (1980)

Highter, W.H., Vallee, R.P.: The liquefaction of different mine tailings under stress-controlled loading. Eng. Geol. **16**, 147–150 (1980)
Leduc, M., Smith, M.E.: Tailings Co-disposal™ innovations for cost savings and liability reduction. The Latin America Mining Record (2003)
Morris, P.H., Williams, D.J.: Co-disposal of washery wastes at Jeebropilly colliery, Queensland, Australia. Trans. Instn. Min. Metall. (Sect. A: Min. Industry) **106** (1997)
Proskin, S., Sego, D., Alostaz, M.: Freeze–thaw and consolidation tests on Suncor mature fine tailings (MFT). Cold Reg. Sci. Technol. **63**, 110–120 (2010)
Qiu, Y., Sego, D.C.: Laboratory properties of mine tailings. Can. Geotech. J. **38**, 183–190 (2001)
Sanin, M.V., Puebla, H., Eldridge, T.: Cyclic behavior of thickened tailings. In: Proceedings of Tailings and Mine Waste 2012, Keyston, Colorado (2012)
Stone, K.J.L., Randolph, M.F., Toh, S., Sales, A.A.: Evaluation of consolidation behavior of mine tailings. J. Geotech. Eng. ASCE **120**, 473–490 (1994)
Taylor, R.N.: Geotechnical Centrifuge Technology. Chapman & Hall, London (1995)
Wickland, B.E., Wilson, G.W.: Self-weight consolidation of mixtures of mine waste rock and tailings. Can. Geotech. J. **42**, 327–339 (2005)
Wickland, B.E., Wilson, G.W., Wijewickreme, D., Klein, B.: Design and evaluation of mixtures of mine waste rock and tailings. Can. Geotech. J. **43**, 928–945 (2006)
Wickland, B.E., Wilson, G.W., Wijewickreme, D.: Hydraulic conductivity and consolidation response of mixtures of mine waste rock and tailings. Can. Geotech. J. **47**, 472–485 (2010)
Wijewickreme, D., Sanin, M.V., Greenaway, G.: Cyclic shear response of fine-grained mine tailings. Can. Geotech. J. **42**, 1408–1421 (2005)
Wilson, G.W., Wickland, B.E., Palkovits, F., Longo, S.: The implementation of paste rock systems for preventing acid rock drainage from waste rock and tailings impoundments. In: Submitted for Oral Presentation at the 8th ICARD, Skelleftea, Sweden, 22–26 June 2009

Probabilistic Assessment of Liquefaction Potential of Guwahati City

Binu Sharma(✉) and Noorjahan Begum

Civil Engineering Department, Assam Engineering College, Guwahati, India
binusharma78@gmail.com, noor2412@gmail.com

Abstract. This paper presents a probabilistic assessment of the liquefaction potential of Guwahati City, in the North Eastern region of India. North Eastern region of India is one of the six most seismically active regions of the world. The assessment was done for saturated cohesionless deposits. Probabilistic liquefaction assessment is done in terms of probability of liquefaction. According to many researchers, probability of liquefaction is a better index for assessment of liquefaction than factor of safety obtained by the deterministic approach. In this papera comparative study of liquefaction potential of soil sites susceptible to liquefaction have been done using eight SPT based probabilistic approaches. Most of the approaches are based on logistic regression analysis of field performance data to determine empirical equations for evaluating the probability of liquefaction. Departing from logistic regression approach other-shave used a Bayesian mapping function that relates reliability index and factor of safety to the probability of liquefaction. Standard penetration test (SPT) N values, engineering properties of the soils and depth of water table were taken from a data base of 200 boreholes up to 30 m depth covering an area of 262 km^2 in Guwahati city. Standard penetration test was done at every 1.5 m interval to determine the N value of the soil with depth. Undisturbed and disturbed soil samples were collected to determine the engineering properties of the soils in the laboratory. A design peak ground acceleration of 0.36 g was used since Guwahati falls in zone V, a high earthquake risk zone, according to the seismic zoning map of India. Liquefaction susceptibility from the methods is presented as maps showing zones of levels of risk of liquefaction. The SPT-based liquefaction evaluation probabilistic procedures have been found to yield significantly different predictions. Comparisons of the methods in the probabilistic approach have shown a difference in the values of probabilities in the same depth. However, on common comparison, the soil layers susceptible to liquefaction in different zones of the city have been identified.

1 Introduction

Guwahati is situated along the bank of river Brahmaputra in the North Eastern (N.E.) region of India. The Indian Standard code of practice identified Northeast India, including Assam as a highly seismic zone by placing it in the highest seismic zonal level i.e. zone V. The Great Assam earthquake of 1897 of magnitude (M) equal to 8.7 caused large scale damage and ground liquefaction in Assam. The Upper Assam earthquake of 1950 of magnitude equal to 8.7 also caused liquefaction in many areas of Assam.

T. Abdoun and S. Elfass (eds.), *Soil Dynamics and Soil-Structure Interaction for Resilient Infrastructure*, Sustainable Civil Infrastructures, DOI 10.1007/978-3-319-63543-9_4

This study is aimed at determining liquefaction potential of Guwahati city by probabilistic approaches. A comparative studyof liquefaction potential of soil sites susceptible to liquefaction have been done using eight SPT based probabilistic approaches. According to many researchers, probability of liquefaction is a better index for assessment of liquefaction than factor of safety obtained by the deterministic approach. Probabilistic versions of Cetin et al. (2004), Boulanger and Idriss (2012), Toprak et al. (1999), Youd and Noble (1997), Juang et al. (2002), Juang et al. (2012) have been used. For the study, data were taken from 200 bore holes of 30 m depth from the ground surface. This data base was from the Department of Science and Technology (DST) Project (2005) completed by the Civil Engineering Department of Assam Engineering College.

2 Background

Toprak et al. (1999) developed SPT-based probabilistic liquefaction boundary curves using logistic regression analyses. The logistic regression equation, obtained from the world wide liquefaction database (total number of data points = 440) is given by the following equation

$$\mathrm{Logit}(P_L) = \ln\left[\frac{P_L}{(1 - P_L)}\right] = 10.4459 - 0.2295(N_1)_{60\,cs} + 4.0573 \ln\left(\frac{CSR}{MSF}\right) \quad (1)$$

Where P_L = probability that liquefaction occurred. In Eq. (1) CSR i.e. cyclic stress ratio, was calculated according to Youd and Idriss (1997). AllCSR were adjusted to M = 7.5 using the Idriss (1999) magnitude scaling factor (MSF). $(N_1)_{60cs}$, which is the corrected equivalent clean-sand blow count, is determined according to Youd and Idriss (1997). Youd and Noble (2001) also used logistic analysis (Eq. 2) to analyze the probability of liquefaction.

$$\mathrm{Logit}(P_L) = \ln\left[\frac{P_L}{(1 - P_L)}\right] = -7.633 + 2.256M_w - 0.258(N_{1_{60CS}}) + 3.095 \ln CRR \quad (2)$$

CRR i.e. cyclic resistance ratio and $(N_1)_{60cs}$ in Eq. (2) are taken from Youd and Idriss (1997) and M_w = moment magnitude of an earthquake.

Juang et al. (2002) performed logistic regression as shown in Eq. (3) on a database of 243 cases with SPT measurements that were taken from a database of field performance cases compiled by Fear and McRoberts (1995).

$$\ln\left[\frac{P_L}{(1 - P_L)}\right] = 10.1129 - 0.2572(N_1)_{60cs} + 3.4825 \ln(CSR_{7.5}) \quad (3)$$

The cyclic stress ratio (CSR) in Eq. (3) is calculated according to Seed et al. (1985). The equivalent clean sand corrected N values i.e. the $(N_1)_{60cs}$values were obtained according to Youd and Idriss (1997).

Juang et al. (2002) proposed the Bayesian mapping approach, on the same database of 243 cases, to evaluate the probability of liquefaction, P_L, as a function of factor of safety, F_S, by using the updated standard penetration test based Seed and Idriss (SPT-SI) method as presented in Youd et al. (2001). This is shown in Eq. (4).

$$P_L = 1/[1 + (F_S/A)^B] \tag{4}$$

Here the regression coefficients, A = 0.77 and B = 3.25, Juang et al. (2012) upgraded computation of probability of liquefaction using standard penetration tests based on an updated database of liquefaction case histories by Idriss and Boulanger (2010). They included various theories, including the principles of Maximum Likelihood and the Information Theory to develop the models of probability. According to them, among the various models, the best is the Gaussian + M_1 model and the Logistic + M_1 model. The logistic + M $_1$ model is given by the following equation

$$P_L = \left[\frac{1}{1 + \exp[7.55(F_s - 0.95)]}\right] \tag{5}$$

Though both the models can be used to determine probability of liquefaction, they considered a mixed model that takes the average of the Gaussian + M_1 model and the Logistic + M_1 model which is shown below: (Eq. 6).

$$P_L = 0.5\left\{1 - \Phi([F_S - 0.95]/0.24) + \frac{1}{1 + \exp[7.55(F_s - 0.95)]}\right\} \tag{6}$$

Where Φ is the standard normal cumulative distribution function. In the above equation, the P_L–F_S mapping function is specific to the Idriss and Boulanger (2004, 2006) method.

Cetin et al. (2004) gave an SPT-based probabilistic relationship which can be expressed by a single, composite relationship which is a function of SPT N value, CSR, moment magnitude (M), Fine content (FC) and vertical effective stress. This relationship is shown in Eq. (7).

$$P_L(N_{1,60}, CSR_{eq}, M_w, \sigma'_v, FC) = \phi\left[-\frac{\left\{N_{1,60}(1 + 0.004FC) - 13.32\ln(CSR_{eq}) - 29.53\ln(M) - 3.70\ln\left(\frac{\sigma'_v}{P_a}\right) + 0.05FC + 16.85\right\}}{2.70}\right] \tag{7}$$

According to Boulanger and Idriss (2012), SPT-based liquefaction triggering probabilistic correlation was derived using a maximum likelihood approach and an updated case history database. After knowing the $CSR_{M=7.5}$,.and $(N_1)_{60cs}$ from the deterministic liquefaction triggering correlation by Idriss and Boulanger (2004, 2006), the corresponding probability of liquefaction relationship was presented as:

$$\begin{aligned}&P_L((N_1)_{60CS}, CSR_{M=7.5,\sigma'_v=1atm})\\&= \phi\left[-\frac{\left\{\frac{(N_1)_{60CS}}{14.1}\right\}+\left\{\frac{(N_1)_{60CS}}{126}\right\}^2-\left\{\frac{(N_1)_{60CS}}{23.6}\right\}^3+\left\{\frac{(N_1)_{60CS}}{25.4}\right\}^4-2.67-\ln\left(CSR_{M=7.5,\sigma'_v=1atm}\right)}{\sigma_{\ln(R)}}\right]\end{aligned} \tag{8}$$

3 Results and Discussions

Asoil database from 200 boreholes was used to determine probability of liquefaction and factors of safety against liquefaction for areas inGuwahati city. For this purpose bore holesof 30 m depth were made in 200 locations covering anarea of 262 km^2. The location of the bore holes are shown in Figs. (6) and (7). Standard penetration test was done at every 1.5 m interval to determine the N value of the soil withdepth. Undisturbed and disturbed soil samples were collected to determine the engineering properties of the soils in the laboratory. Guwahati soil is mostly alluvial in nature. Guwahati soil mostly consists of alternating layers of both fined grained and coarse grained soils. There is a great deal of variation in thethickness of these layers. The fine grained fraction mostly consists of red, brown and gray colored siltyclay and clay of classification CL and CH according to USCS. In a few locations inorganic silt of classification ML and CL–ML and non plastic inorganic silts were also encountered. The soils are mostly of classification SP, SW, SC, SM, SP-SC. Gravel deposits were also encountered in certain boreholes. The 200 boring logs showed the water tableto be within 0–6 m meter of the ground surface. The map showing depth to ground water level in Guwahati city is shownin Sharma and Hazarika (2013). The map indicates that in most ofthe locations in Guwahati city the water table is at ashallow depth. The water level fluctuation data obtained for the period 2001–2010 by the Central Ground Water Commission, India, shows the waterlevel rise to be less than 2 m and the fall of water levelis also less than 2 m in Assam. In this present study the soil layers that were identified for liquefaction analysis are fine to medium sand and silty sands that have classification of SP, SW, SC, SM, SP-SC. Inorganic silt of classification ML, ML-CL and non plastic inorganic silts were also analysed for liquefaction susceptibility. Evaluation of liquefaction potential requires peak ground acceleration (a_{max}) during earthquake. For its estimation ground motion relation are necessary where a_{max} is expressed as a function of magnitude and rupture distance. No such relation is available for N.E. India. In the absence of such a relation, the peakground acceleration, $a_{max,}$ for Guwahati city is taken as 0.36 g. This is according to IS 1893 (2002) which puts Guwahati in Zone V, a severe seismic zone. This is for an 8.1 magnitude earthquake occurring on a fault at an epicentral distance of 50 km from Guwahati city.

Determination of probability of liquefaction of Guwahati city has been done by the above mentioned probabilistic procedures. Of the 200 sites, 49 sites in Guwahati have been found to be susceptible to liquefaction according to the Toprak et al. (1999), Youd and Noble (1997) methods of probability analysis and 50, according to Juang et al. (2002), Idriss and Boulanger (2012), Juang et al. (2012) methods of analysis. According to Cetin et al. (2004), 51 sites in Guwahati have been found to be susceptible to liquefaction. The rest of the sites where the bore holes are located are not susceptible to liquefaction.

Comparison of the probability values by the eight different approaches for Boreholes 7 (Azara), 8 (Azara,Vet.), 12 (VIP Road), 16 (Lower Mirjapur village) and 43 (Aathgaon) for soil layers susceptible to liquefaction for an earthquake magnitude 7.5 are shown from Fig. 1(a) to (e).

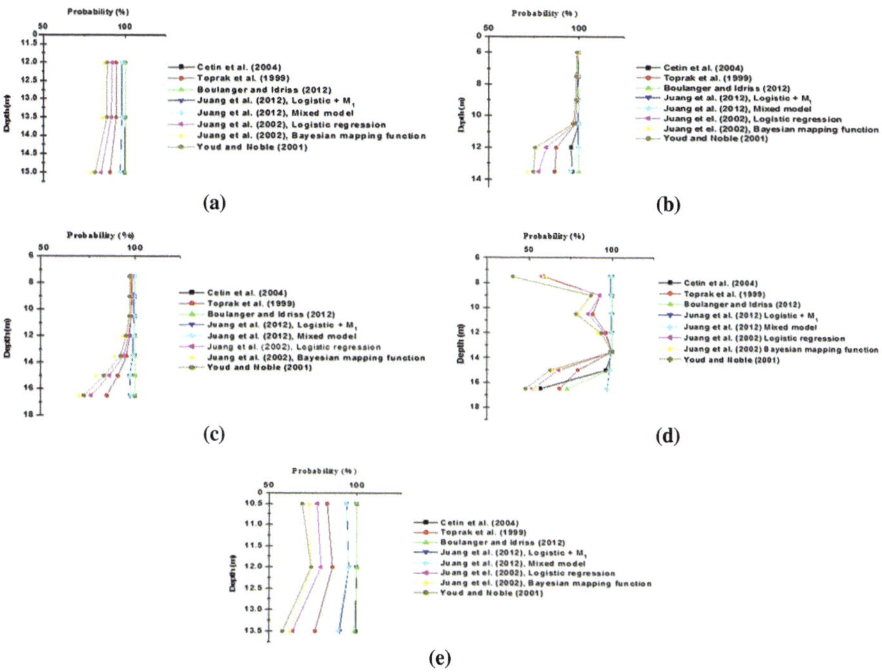

Fig. 1. (a). Comparison of probability of liquefaction for borehole 7, (b). Comparison of probability of liquefaction for borehole 8, (c). Comparison of probability of liquefaction for borehole 12, (d). Comparison of probability of liquefaction for borehole 16, (e). Comparison of probability of liquefaction for borehole 43

The above five figures are selected arbitrarily. Boreholes 7, 8, 12, 16 and 43 show the same pattern of variability. Higher probability values are obtained by Idriss and Boulanger (2012), Juang et al. (2012), Cetin (2004) and lower values are shown by Youd and Noble (1997), Juang et al. (2002) Bayesian mapping. Toprak et al. (1999) gives probability values that lie in between Boulanger and Idriss (2012), Juang et al. (2012), Cetin (2004), Youd and Noble (1997), Juang et al. (2002). However this pattern is not observed in a few bore holes. Logistic regression gives higher values of probability than the Bayesian mapping approach. The logistic + M_1model and the mixed model give the same values of probability.

There is variation between the probability values at the same depth by the different approaches. The average percentage variation of all the other methods with respect to Boulanger and Idriss (2012) for the Boreholes susceptible to liquefaction for an earthquake magnitude 7.5 were calculated. This are shown in Table 1.

Table 1. Percentage variation of the different probability models with respect to Idriss and Boulanger (2012)

Toprak et al. (1999)	Cetin et al. (2004)	Juang (2002) Logistic regression	Juang (2002) Bayesian mapping	Youd and Noble (2001)	Juang (2012) (Logistic $+M_1$)model	Juang (2012) Mixed model
20.09	16.55	21.96	24.85	24.81	10.68	10.36

The variation is less in case of Juang et al. (2012) because the same updated data set of Idriss and Boulanger (2010) have been used by Juang (2012) and the factor of safety of Idriss and Boulanger (2004) have been used by them to developed the probabilistic models. The database by Idriss and Boulanger (2010)consists of 227 cases, out of which 115 are nonliquefied cases and 112 are liquefied cases.

According to Juang et al. (2012),the model developed and recommended by Boulanger and Idriss (2012) for practice can be expressed by Eq. (9) after re-arrangement.

$$P_L(F.O.S) = 1 - \phi\left[\frac{\ln(F_S) + 0.13}{0.13}\right] \tag{9}$$

It is noted that the liquefaction probability is expressed as a function of the factor of safety. The probabilityversus factor of safety relationship for 50 Boreholes is shown in Fig. (2) where it is seen that liquefaction probability is presented as a function of factor of safety. The lowest factor of safety against liquefaction and the lowest values of probability of liquefaction, among the layers, is retained as the factor of safety and probability for that Borehole.

Figure (3) depicts the P_L - F_S relationship according to Juang et al. (2002) with probability calculated from logistic regression analysis and from the Bayesian mapping approach. In the Bayesian mapping approach, probability is a direct function of factor of safety. In the logistic regression approach the formulation of the equation is different, however a similar trend as for the Bayesian mapping approach is observed.

According to Juang et al. (2012), liquefaction probability can be calculated by the logistic + M_1 and the mixed model. Liquefaction probability is expressed as a function of the factor of safetyin both the models. Therefore in Fig. (4) probability versus factor of safety is shown as a smooth curve and the results of both the models agree with one another. The models by Juang et al. (2012), employed the same database of 227 cases developed by Idriss and Boulanger (2010). The database developed by Idriss and Boulanger (2010) is regarded as an update of the Cetin et al. (2004) database, although different opinions exist for some cases.

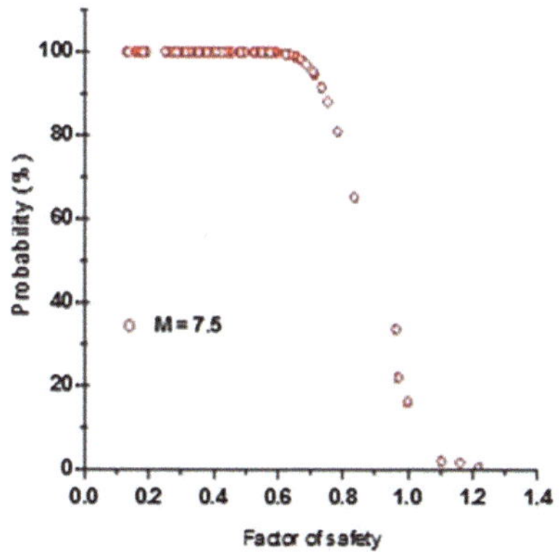

Fig. 2. Relationship between probability by Boulanger and Idriss (2012) and factor of safety by Idriss and Boulanger (2004)

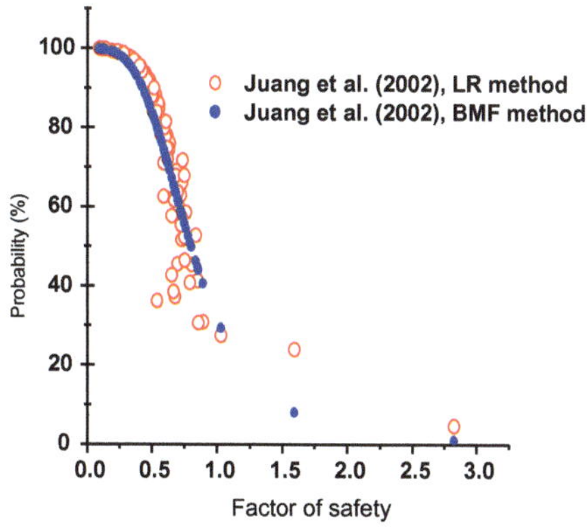

Fig. 3. Relationship between probability and factor of safety by Juang et al. (2002)

Figure (5) shows the same relationship according to the Cetin et al. (2004) model. It is noted that the liquefaction probability is not a function of the safety factor in the Cetin et al. (2004) model because there is no obvious definition of factor of safety in the model. The probabilistic approaches of Youd and Nobel (1997), Toprak et al. (1999) are based on earlier and less comprehensive databases and the relationships between probability and factor of safety are not discussed here.

The liquefaction potential map of Guwahati city for an earthquake magnitude of 7.5, according to the probabilistic assessment of Juang et al. (2012) mixed model is shown in Fig. (6) and according to Juang et al. (2002) is shown in Fig. (7).

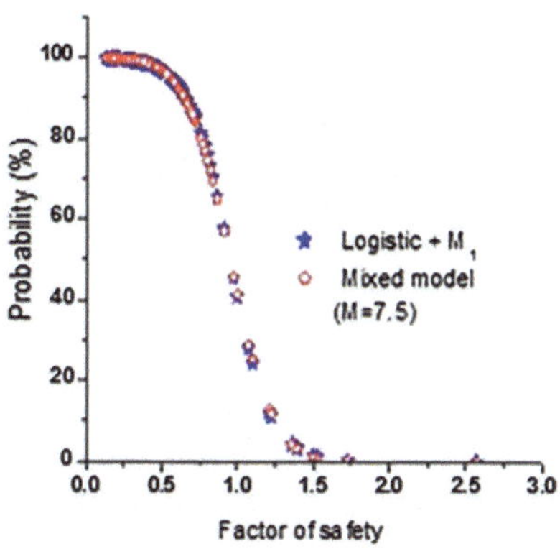

Fig. 4. Relationship between probability by Juang et al. (2012)and factor of safety by Idriss and Boulanger (2004)

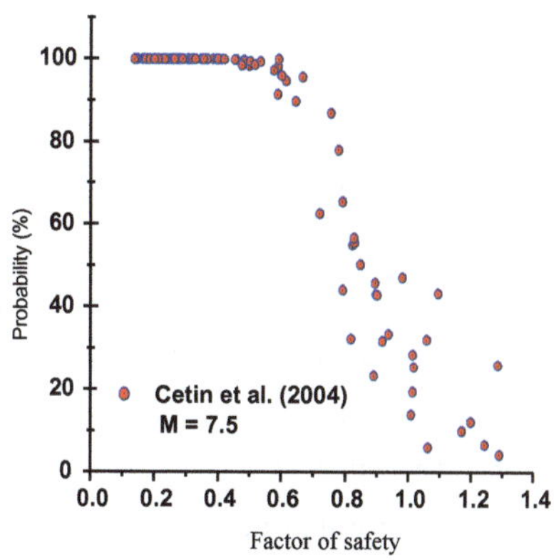

Fig. 5. Relationship between probability and factor of safety by Cetin et al. (2004).

In developing the maps the lowest values of probability of liquefaction, among the layers, is retained as the probability of liquefaction for that Borehole. The maps show zones of different levels of risk of liquefaction. It is observed from the maps that the southern bank of river Brahmaputra with the areas of Azara, Guwahati Airport area, Jhalukbari, Pandu, Bharalumukh and Uzanbazar, some areas in G.S.Road, Gorchuk area, PachimBoragaon and areas near Chandmari, Narengi and Hatigaon are most susceptible to liquefaction. The northern bank of the city is also susceptible to liquefaction. The above two maps are chosen as representing the Bayesian mapping approach and Juang et al. (2012) is based on the updated databases of Idriss and Boulanger (2010).

In the southwest zone covering Azara, Guwahati Airport area, Jhalukbari and Pandu area, the liquefiable layers extend from 6 m to around 16.5 m from the existing ground surface. In the northern zone in Abhaypur location the liquefiable layer extend from 10.5 m to 15 m from the ground surface and in the other liquefiable locations, liquefaction has been manifested only at one or two particular depths. Similarly in the southeast zone a few bore holes in Chandmari, Narengi, Kahilipara, Hatigaon and Beltolamouza have been found to be susceptible to liquefaction with liquefaction manifested at one or two particular depths. The possible zone of liquefaction usually extends from the ground surface to maximum depth of about 15 m. However in this study depth of liquefaction has been observed up to 16.5 m.

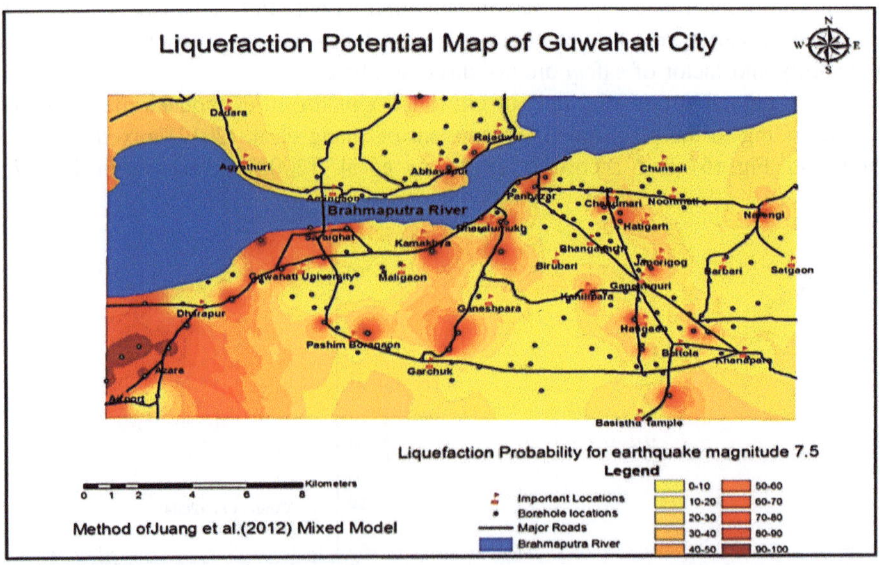

Fig. 6. Probabilistic liquefaction potential map of Guwahati city (Juang et al. 2012)

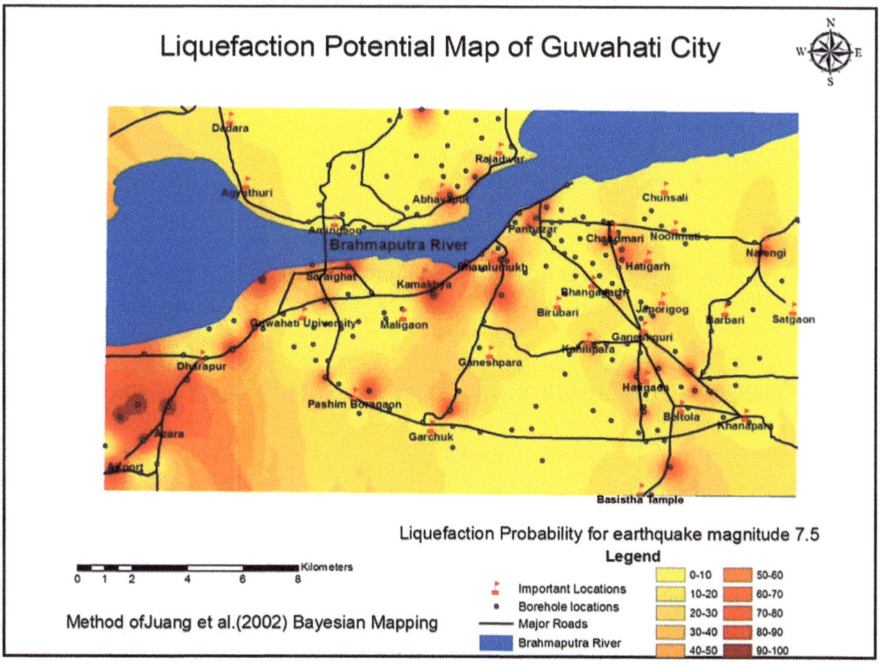

Fig. 7. Probabilistic liquefaction potential map of Guwahati city (Juang et al. 2002)

4 Conclusions

A comparative studyof liquefaction potential of soil sites susceptible to liquefaction has been done using eight SPT based probabilistic approaches. Higher probability values were obtained by Boulanger and Idriss (2012), Juang et al. (2012), Cetin (2004) and lower values were shown by Youd and Noble (1997), Juang et al. (2002). Toprak et al. (1999) gave probability values in between Idriss and Boulanger (2012), Juang et al. (2012), Cetin (2004), Youd and Noble (1997), Juang et al. (2002). However this pattern was not observed in a few bore holes. There was variation between the probability values at the same depth by the different approaches. GIS based probabilistic liquefaction map of Guwahati city show that the southern bank of river Brahmaputra with the areas of Palashbari, Airport area, Azara, Jhalukbari, Pandu, Bharalumukh and Uzanbazar, some areas in G.S.Road, Gorchuk area and areas near Chandmari have probabilities of liquefaction greater than 50%. Some areas in the northern bank of the city are also susceptible to liquefaction. In the southwest zone, the liquefiable layers extend from 6 m to around 16.5 m from the existing ground surface. In the northern zone in Abhaypur location the liquefiable layers extend from 10.5 m to 15 m from the ground surface and in the other liquefiable locations, liquefaction has been manifested only at one or two particular depths.

Acknowledgment. The Geotechnical data of the 200 boreholes were taken from a project work given to Assam Engineering College, titled "Liquefaction potential determination of Guwahati city" funded by The Department of Science and Technology, India for Microzonation of Guwahati city. We acknowledge the help and assistance given by DST, India for the study.

References

Boulanger, R.W., Idriss, I.M.: Probabilistic standard penetration test based liquefaction triggering procedure. J. Geotech. Geoenviron. Eng. **138**, 1185–1195 (2012). doi:10.1061/(ASCE)GT.1943-5606.0000700

Cetin, K.O. et al.: Standard penetration test-based probabilistic and deterministic assessment of seismic soil liquefaction potential. J. Geotech. Geoenviron. Eng. (2004). doi:10.1061/(ASCE)1090-0241(2004)130:12(1314)

Fear, C.E., McRoberts E.C.: Report on liquefaction potential and catalogue of case records. Internal Res. Rep., Department of Civil Engineering, University of Alberta, Edmonton, Alberta, Canada (1995)

Idriss, I.M.: An update of the Seed–Idriss simplified procedure for evaluating liquefaction potential. In: Proceedings of Transportation Research Board Workshop on New Approaches to Liquefaction Analysis (1999)

Idriss, I.M., Boulanger, R.W.: Semi-empirical procedures for evaluating liquefaction potential during earthquakes. In: 11th International Conference on Soil Dynamics & Earthquake Engineering (ICSDEE) and the 3rd International Conference on Earthquake Geotechnical Engineering (ICEGE) (2004)

Idriss, I.M., Boulanger, R.W.: Semi-empirical procedures for evaluating liquefaction potential during earthquakes. Soil Dyn. Earthq. Eng. (2006). doi:10.1016/j.soildyn.2004.11.023

Idriss, I.M., Boulanger, R.W.: Report on SPT-based liquefaction triggering procedures. Center for geotechnical modeling, Department of civil and environmental engineering, University of California at Davis (2010)

Juang, C.H. et al.: Reliability-based method forassessing liquefaction potential of soils. J. Geotech. Geoenviron. Eng. (1999) doi:10.1061/(ASCE)1090-0241(1999)125:8(684)

Juang, C.H., Jiang, T.: Assessing probabilistic methods for liquefaction potential evaluation. Soil Dyn. Liquefaction (2000). doi:10.1061/40520(295)10

Juang, C.H., et al.: Assessing probability-based methods for liquefaction evaluation. J. Geotechn. Geoenviron. Eng. (2002). doi:10.1061/(ASCE)1090-0241(2002)128:7(580)

Juang, C.H., et al.: New models for probability of liquefaction using standard penetration tests based on an updated database of case histories. Eng. Geol. J. (2012). doi:10.1016/j.enggeo.2012.02.015

Seed, H.B., et al.: The influence of SPT procedures in soil liquefaction resistance evaluations. J. Geotech. Eng., 1425–1445 (1985). doi:10.1061/(ASCE)0733-9410(1985)111:12(1425)

Sharma, B., Hazarika, P.: Assessment of liquefaction potential of Guwahati city. a case study. Int. J. Geotech. Geol. Eng. (2013). doi:10.1007/s10706-013-9667-x

Toprak, S., et al.: CPT and SPT-based probabilistic assessment of liquefaction potential. In: 7th U.S.–Japan Workshop on Earthquake Resistant Design of Lifeline Facilities and Countermeasures Against Liquefaction, Seattle (1999)

Yegian, M.K., Whitman, R.V.: Risk analysis for ground failure by liquefaction. J. Geotech. Eng. **104**, 921–938 (1978)

Youd, T.L., Noble, S.K.: Liquefaction criteria based on statistical and probabilistic analyses. In: NCEER Workshop onEvaluation of Liquefaction Resistance of Soils, NCEER TechnicalRep. No: NCEER-97-0022, pp. 201–205 (1997)
Youd, T.L., Idriss, I.M.: NCEER Workshop on Evaluation of Liquefaction Resistance of Soils, Technical report No. NCEER-97-0022. State Univ. of New York at Buffalo, Buffalo (1997)
Youd, T.L., et al.: Liquefaction resistance of soils: Summary report from the 1996 NCEER and 1998 NCEER/NSF workshops on evaluation of liquefaction resistance of soils. J. Geotech. Geoenviron. Eng. **127**, 817–833 (2001)
IS (1893–2002) Part-1. Indian standard criteria for earthquake resistance design of structures. BIS, New Delhi

Lateral Response of Socketed Pile Under Cyclic Load

Annamalai Rangasamy Prakash(✉) and Kasinathan Muthukkumaran

National Institute of Technology, Tiruchirappalli, Tamil Nadu, India
prakashartn@gmail.com, kmk@nitt.edu

Abstract. The pile, carrying large axial and lateral load is often subjected to lateral cyclic load due to the wave action of water, glacier movement and by heavy wind action. Transfer of heavy load requires the pile to socket into the hard stratum. The existing p-y curve for socketed pile gives an approximate solution under static load. In order to overcome this, a series of experimental program was conducted on a model pile by varying its length of embedment to the length of pile ratio (L_e/L) 25%, 50% and 75%. The model pile of length 1200 mm was made up of a hollow aluminium pipe of 25.4 mm outer diameter and 3 mm thickness. Electrical resistance foil type strain gauges of 350 Ω were placed along the length of pile at equal intervals of 100 mm. LVDT's were placed on the pile head and ground level to measure the lateral deflection. Lateral load was applied gradually at the pile head, the strains and deflection were measured using a data acquisition system connected to a computer. Load deformation response for varying frequencies and cycles were generated on the basis of the experimental analysis. The experimental results helped the observation that, socketing the pile has significant effects on lateral load carrying capacity. Embedding the pile into hard strata reduces the lateral displacements substantially compared with pile bearing over hard strata.

Keywords: Pile socketing · Experimental program · Model pile instrumentation · Embedded length · Lateral deflection response

1 Introduction

Pile can transfer a large volume of load from superstructure such as tall structures, offshore/onshore wind structures, elevated express ways, bridge piers/abutments in loose sand layer overlying hard strata. The lateral load carrying capacity can be further increased by socketing the pile shaft into a hard stratum. The lateral load carrying capacity of socketed pile depends on depth of socket and the ultimate soil resistance offered by the soil and overlaying hard strata. However, in addition to static lateral loading, wind, waves, earth pressures, and water pressures may place cyclic lateral loads on pile-supported structures, as described by Long et al. (1994). A number of researchers, including Prakash (1962), Broms (1964), Davisson and Salley (1968), Alizadeh and Davisson (1970), and Davisson (1970), have suggested a reduction in the value of the static coefficient of soil reaction to take into account the effects of cyclic lateral loads on piles. Degraded static p-y curves have been proposed by Reese et al.

T. Abdoun and S. Elfass (eds.), *Soil Dynamics and Soil-Structure Interaction for Resilient Infrastructure*, Sustainable Civil Infrastructures, DOI 10.1007/978-3-319-63543-9_5

(1974), based on some full-scale cyclic pile load test results. Cyclic p-y curves obtained by reducing the static soil resistance were constructed by Little and Briaud (1988), on the basis of in situ pressuremeter test results. A more detailed and systematic study on the effects of cyclic lateral loads on piles in sand was conducted by Long and Vanneste (1994) using a linearly increasing soil-reaction modulus and degradation of the static p-y curve. Hence, a series of 1 g experimental program was carried out to overcome this problem.

2 Experimental Program

2.1 Model Pile Instrumentation

A model pile of 25.4 mm diameter with L/D ratio 47.24 made of aluminium of type 6061-T6 conforming to ASTM-B241 has been used in this study. Table 1 shows the geometric and physical details of the model pile. The model piles were instrumented with electrical resistance foil type strain gauges of resistance 350 Ω. A total of 12 strain gauges, as shown in Fig. 1, were pasted along the length of the pile at equal intervals of 100 mm and width of the strain gauge was 5 mm with a gauge factor of 1.9. The strain gauges were coated with silicon to ensure protection from external contact. The lead wire from the gauges was taken into the pile by making a tiny hole just above the position of the strain gauge and collected at the pile top. The wires from the strain gauges were connected to a Data Acquisition system forming the Wheatstone bridge circuit for measuring the resistance offered by the pile to the respective load. A simple cantilever test was conducted for calibrating the strain gauges for bending moment along the pile length (Kong et al.). The strain response was found to be linear and a bending constant of 76.01 Nmm was arrived at from the strain response curve. The bending moment for the respective strain value can be obtained from the Eq. (1)

$$M = C \times \varepsilon \tag{1}$$

Where, M is the Bending moment, C is the bending constant (76.01 Nmm) and ε is the measured strain value.

Table 1. Geometric and engineering properties

Parameters	Values
Outer Diameter OD (mm)	25.4
Inner Diameter ID (mm)	19.4
Model pile length (mm)	1200
Young's modulus E_p (GPa)	69
Flexural stiffness E_pI_p (Nmm^2)	9.30E + 08
L/D ratio	47.24

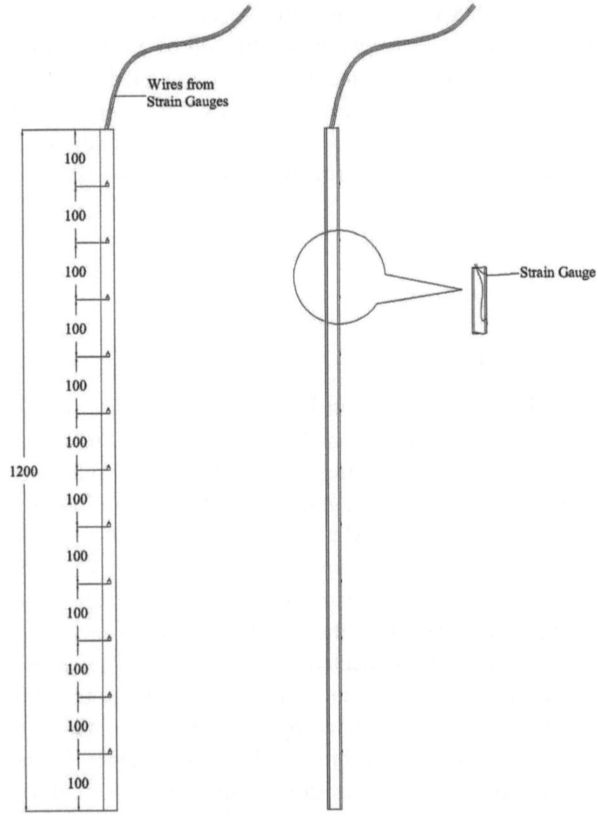

Fig. 1. Instrumented model pile

2.2 Soil Sample Preparation

Dry river sand used in the study was collected from the nearby river Cauvery and classified as poorly graded sand (SP) as per IS 1498/ASTM D-2487. Figure 2 shows details of the gradation of poorly graded sand. The details of index and engineering properties are given in Table 2. Relative density experiments were carried out as per IS 2720- part (14)/ASTM 1425 and the maximum and the minimum densities values were found to be 19.61 kN/m^3 and 14.96 kN/m^3 respectively. Figure 3 shows the microscopic image of the poorly graded sand and the particles in the sub-rounded shape, that helped in maintaining uniform density in filling and the angle of internal friction was maintained throughout the experiments as stated by Lee et al.(2006). Direct shear test was conducted as per IS 2720-Part (13) for a density of sand in shear box maintained at an uniform rate of 1.49 g/cc and values are found to be 34.16°.

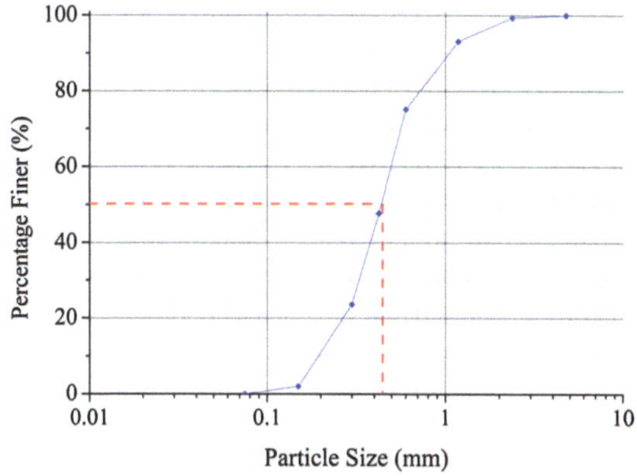

Fig. 2. Gradation of poorly graded sand

Table 2. Details of index and engineering properties of soil sample

Engineering property	Outer Diameter OD (mm)
Specific gravity (Gs)	2.62
Effective diameter (D_{10})	0.2 mm
Coefficient of uniformity (C_u)	2.56
Coefficient of curvature (C_c)	1.16
Average particle (D_{50})	0.43 mm
Max. dry density (ρ_{max})	1.96 g/cc
Min. dry density (ρ_{min})	1.49 g/cc

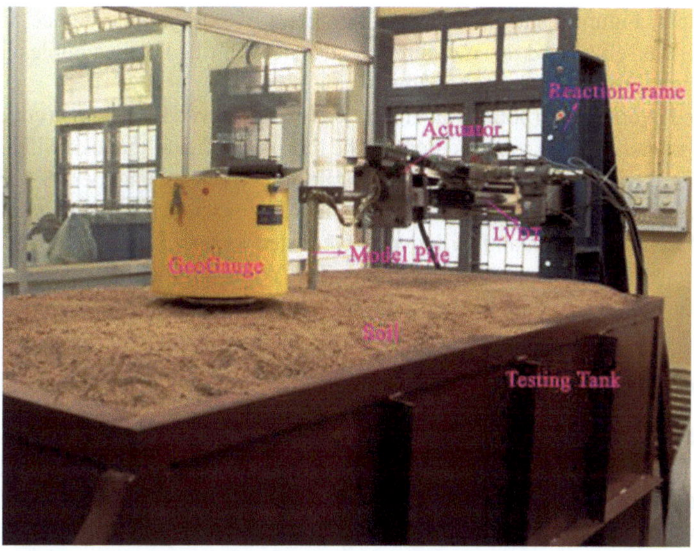

Fig. 3. Experimental setup

2.3 Testing Chamber

A rectangular steel tank with internal dimensions of 2 m × 1 m × 1 m, length, width and height was designed and fabricated for minimizing the boundary effects caused while the lateral load acted on the pile head. The distance from the center of the model pile to the edge of the testing tank in loading direction was 39D, is greater than 8D where the effect was smaller as suggested by Gandhi et al. (1997), Muthukkumaran et al. (2014). The steel testing tank was kept at the axis perpendicular to the reaction frame holding the loading arrangements as shown in Fig. 3.

2.4 Testing Procedure

A model pile was installed as pre-driven method, while the pile was kept in a loading position before start of the filling of the soil sample with the help of a guiding rod. C-clamps were employed to keep the pile in a vertical position and removed during loading. The sand raining method described by Gandhi et al.(1997), Muthukkumaran et al. (2014) was used in this study for achieving a uniform density in filling. This can be achieved by maintaining the constant falling height of sand particles from sand raining device. Figure 4 shows the sand raining device containing a cylindrical hopper of diameter 180 mm and of height 150 mm connected to a long funnel of diameter 25.4 mm to a length 750 mm. Sand particles passing through the long funnel provide a uniform rate of falling under gravity filling, an inverted cone of angle 60° was kept at the bottom of funnel for even disposal of the sand particles. The sand raining device was calibrated to varying height of fall by conducting the density index test as per IS 2720-Part (13), in order to achieve a relative density (R_D) of 30%, 50% and 80% showing the loose, medium and dense states of sand. In this study, a relative density of (R_D) 30% was maintained for all sets of experiments, achieved by keeping the height of fall as 50 mm. Figure 5 shows the calibration chart of sand raining device. Geo-gauge equipment was employed to measure the density of filled soil samples at regular interval heights for all sets of experiments.

The sand overlie hard (bottom layer) stratum was achieved by casting a concrete layer around the model pile to a required depth. The hard stratum was made of M50 mix grade as per IS 10262-2009 to classify the hard stratum as weak rock as per IS 13365-2005. Care was taken during grouting of concreting in order to ensure homogeneity of mix. The strength of concrete cubes and cylinders were tested as per IS 516-2008 and the compressive strength was found to be 52 N/mm^2. The Young's modulus of synthetic rock is 36228 N/mm^2 which resembles the modulus value soft rock.

2.5 Testing Program

A total of 30 experiments as shown in Table 3, were carried out by varying the frequency of loading range from 0.1 Hz to 1.5 Hz and loading cycles from 10 to 500. Lateral load tests were conducted by a displacement controlled actuator of capacity 10 kN. The load was applied at the pile head level. LVDT's of travel length 50 mm with accuracy of 0.01 mm were placed at a regular interval of 200 mm along the pile length up to the ground level to measure the magnitude of pile deflection.

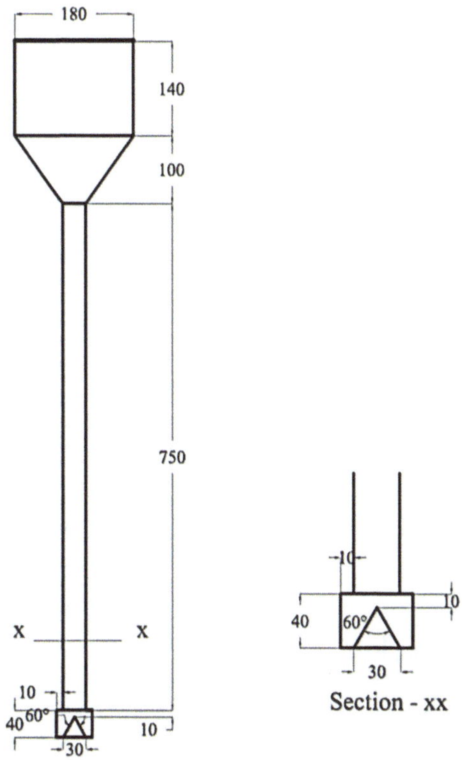

Fig. 4. Sand raining device

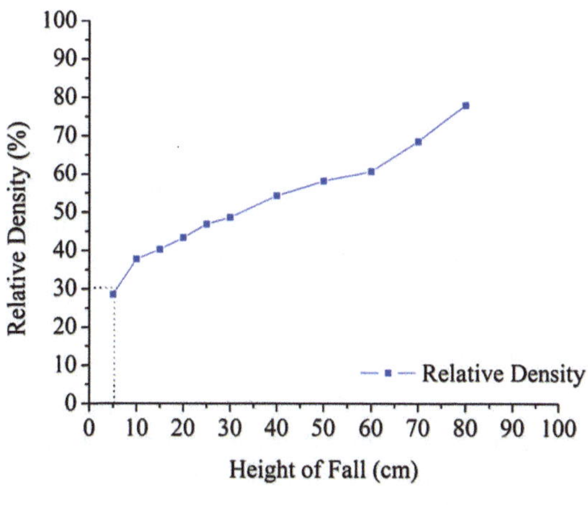

Fig. 5. Calibration of sand raining device

The resistance to deformation of pile head was measured with the help of load cell attached to the actuator with an accuracy 0.01 N. The readings from LVDT's, load cell and strain gauges were recorded with the help of the computer controlled Data Acquisition System. The torsional rotation at the pile head was measured with the help of a torque transducer and found to be zero.

Table 3. Experimental program

Length of Embedment to total pile length ration L_e/L	Loading cycles	Frequency (Hz)
0.75, 0.5 and 0.25	10	0.1
	10	0.25
	10	0.5
	10	0.75
	10	1
	10	1.5
	25	1.5
	50	1.5
	100	1.5
	500	1.5

3 Result and Discussion

3.1 Classification of Model Piles

The ultimate lateral resistance of the pile depends on pile behaviour. Broms (1964) equation classifies the pile behaviour on the basis of the soil subgrade reaction and the flexural stiffness (EI) of the pile. Unsaturated cohesionless sand has been used in the present study; where soil modulus linearly varies with depth and is minimum at the ground level.

$$T = \left(\frac{EI}{\eta_h}\right)^{\frac{1}{5}} \quad (2)$$

η_h = coefficient of modulus of subgrade reaction. Table 1 shows the behaviour of experimental piles, for the pile having minimum free standing height (L_e = 300 mm) of 25.4 mm diameter pile behaving as a long flexible pile. The remaining model pile has free standing height (L_e) of 600 mm and 900 mm behave as a short rigid pile.

3.2 Cyclic Lateral Load-Pile Response

The lateral load capacity of long flexible pile depends on the permissible deflection at the ground level. Figure 6 presents the experimentally measured cyclic lateral loading response of model pile having less free standing height (L_f = 300 mm) and maximum embedded depth of pile in soil and hard stratum (L_e = 900 mm) of L_e/L ratio 0.75 for the frequency of 1.5 Hz. The experimental result shows increase in number of loading not having any significant effect in deflection response of pile as shown in Fig. 7. In the case of increase in loading frequency from 0.1 Hz to 1.5 Hz (mild environmental condition to moderate condition) the pile defection was more with initial loading cycles and got decreased. The Figs. 8 and 9 shows the load carrying capacity for 10 cycles and 50 cycles respectively for a constant frequency of 0.25 Hz with 5 mm amplitude. In the case of pile load carrying capacity, the increase in loading rate has a significant influence in load carrying capacity of pile. This is due the stiffness degradation of the soil around the pile

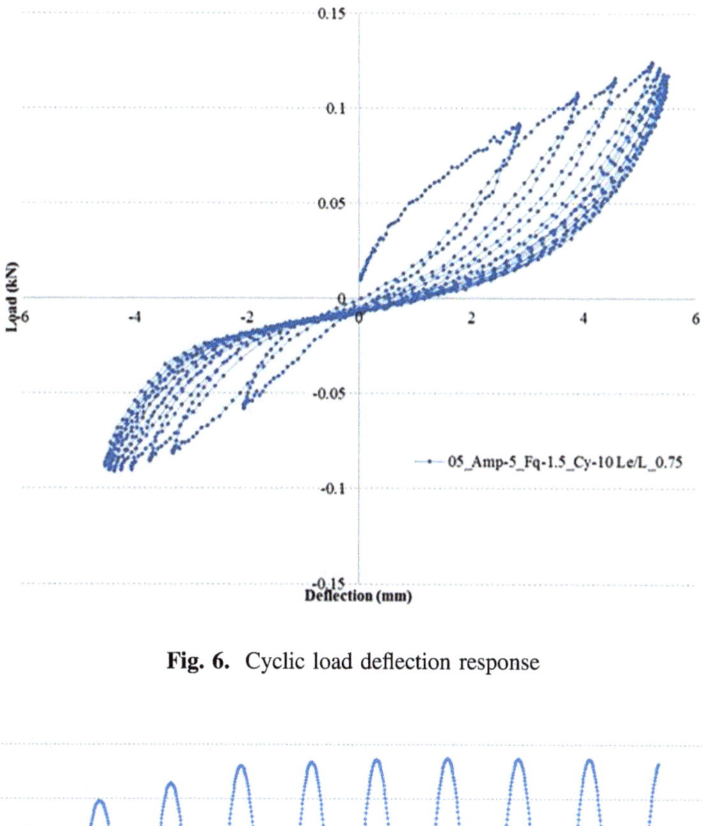

Fig. 6. Cyclic load deflection response

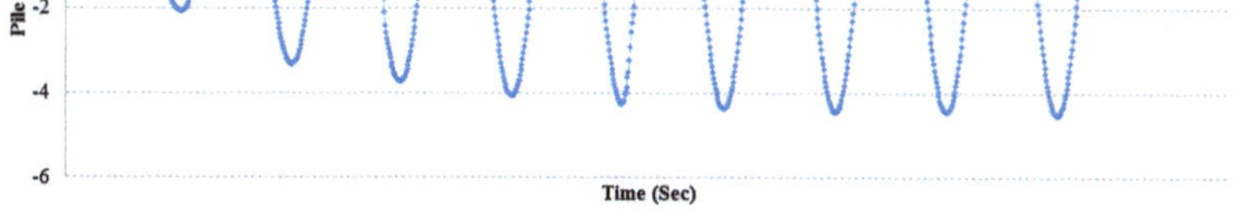

Fig. 7. Cyclic load deflection response

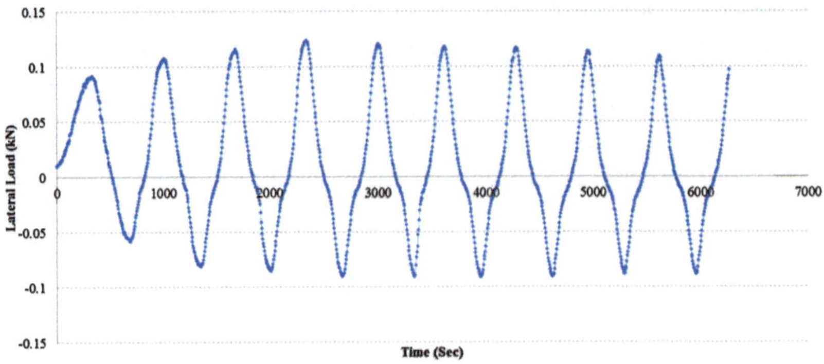

Fig. 8. Cyclic load deflection response

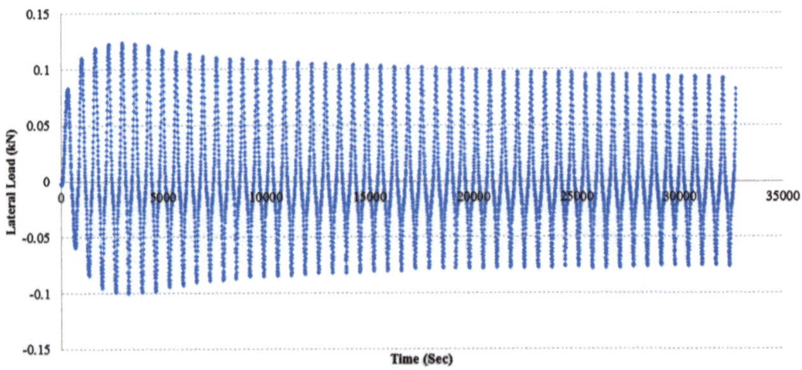

Fig. 9. Cyclic load deflection response

4 Conclusion

The cyclic lateral behaviour of socketed monopile pile has been investigated in this paper by conducting a set of 30 1 g experimental studies on instrumented model pile having 25.4 mm diameter socketed pile. From the experimental studies the following conclusions are drawn.

Increase in the rate of loading does not have any significant effect in the deflection response of the pile.

In the case of the pile load carrying capacity, increase in loading rate has a significant influence in load carrying capacity of the pile. This is due the stiffness degradation of the soil around the pile.

List of notation

The following symbols are used in this paper

ε = Bending strain
Φ = Angle of internal friction
C = Bending constant
C_c = Co-efficient of Curvature
C_u = Co-efficient of Uniformity
D = Diameter of the pile
D_{10} = Percentage of particle size finer than 10%
D_{50} = Percentage of particle size finer than 50%
DAS = Data acquisition system
G_s = Specific Gravity
I_p = Moment of inertia pile
K = Gauge factor
L = Length of pile
L_e = Embedment length of pile in soil and hard stratum
L_f = Free standing length of pile above the ground level

References

Alizadeh, M., Davisson, M.T.: Lateral load test on piles - Arkansas River project. J. Soil Mech. and Found. Engg. Div. ASCE **96**(5), 1583–1604 (1970)

ASTM: Standard practice for classification of soils for engineering purposes (Unified Soil Classification System). D2487-06 (2006)

Broms, B.B.: Lateral resistance of piles in cohesion-less soils. J. Soils Mech. Found. Div. ASCE **90**(2), 27–63 (1964)

Carter, J.P., Kulhawy, F.H.: Analysis of laterally loaded shafts in rock. Geotech. Eng. **118**(6), 839–855 (1992)

Davisson, M.T., Salley, J.R.: Lateral load tests on drilled piers. In: ASTM Symposium on Deep Foundations, pp. 68–83. ASTM, West Conshohocken (1968)

Gandhi, S.R., Selvam, S.: Group effect on driven piles under lateral load. J. Geotech. Geoenviron. Eng. **123**(8), 702–709 (1997)

IS 14593-1998: Design and construction of bored cast-in-situ piles founded on rocks – Guidelines. Bureau of Indian Standards, New Delhi (2003)

Yang, K.: Analysis of laterally loaded drilled shafts in rock. Ph.D. thesis, Univ of Akron (2006)

Little, R.L., Briaud, J.L.: Full scale cyclic lateral load tests on six single pile in sand. Misc. Paper GL-88-27, Geotechnical Div., Texas A&M University, College Station, Tex (1988)

Liang, R., et al.: p-y criterion for rock mass. J. Geotech. Geo-environ. Eng. **135**(1), 26–36 (2009)

Long, J.H., Vanneste, G.: Effects of cyclic lateral loads on piles in sand. J. Geotech. Engrg. ASCE **120**(1), 225–244 (1994)

Matlock, H., Reese, L.C.: Generalized solutions for laterally loaded pile. Trans. ASCE **127**, 1220–1247 (1962)

Muthukkumaran, K.: Effect of slope and loading direction on laterally loaded piles in cohesionless soil. Int. J. Geomech. ASCE **14**(1), 1–7 (2014)

National Cooperative Highway Research Program: Rock-socketed shafts for highway structure foundations. NCHRP synthesis 360 (2006)

Prakash, S., Kumar, S.: Nonlinear lateral pile prediction in sand. J. Geotech. Eng. **122**(2), 130–138 (1996)

Prakash, S.: Behavior of pile groups subjected to lateral loads. Ph.D. dissertation, University of Illinois, Urbana, Ill (1962)

Reese, L.C.: Analysis of laterally loaded piles in weak rock. J. Geotech. Geo-environ. Eng. **123** (11), 1010–1017 (1997)

Uncuoğlu, E., Laman, M.: Lateral resistance of a short rigid pile in a two-layer cohesionless soil. ACTA Geotech. Slov. **2**, 19–43 (2011)

Zhang, L., et al.: Nonlinear analysis of laterally loaded rock-socketed shafts. J. Geotech. Geo-environ. Eng. **126**(11), 955–968 (2000)

Numerical Analysis of Liquefaction Susceptibility of Reinforced Soil with Stone Columns

Zeineb Ben Salem(✉), Wissem Frikha, and Mounir Bouassida

LR14ES03-Ingénierie Géotechnique, Ecole Nationale d'Ingénieurs de Tunis, Université de Tunis El Manar, BP 37 Le Belvédère, 1002 Tunis, Tunisia
Zeineb.bensalem@gmail.com, Frikha_wissem@yahoo.fr, Mounir.bouassida@fulbrightmail.org

Abstract. The present paper studies the use of stone columns as a means of mitigating liquefaction potential. A numerical model using the 2D finite difference code FLAC is developed and validated with published experimental results from literature to simulate the shaking table test. The liquefaction potential of silty sand layer that was improved with stone columns was evaluated in terms of the maximum pore water pressure ratio r_u which is usually defined as the ratio of pore water pressure to the initial vertical effective stress in the soil. Soil liquefaction potential has often been defined when r_u reaches 1.0. Numerical analyses showed that the liquefaction potential decreased after stone columns reinforcement. Parametric analyses were performed to investigate the effect of the excitation frequency and amplitude and the effect of stone columns installation process on liquefaction potential.

Keywords: Liquefaction · FLAC · Shaking table · Stone columns · Reinforced soil

1 Introduction

Investigation of soil liquefaction is one of the main subjects in recent decades. Various ground improvement methods were used, including densification, reinforcement, grouting/mixing, and drainage to reduce the risk of liquefaction potential and its associated damages (Mitchell 2008). Reinforcement with granular columns is the most widely adopted method to mitigate liquefaction (Mitchell and Wentz 1991). Various analytical approaches were proposed to study the performance of reinforcement with stone columns as a method of mitigating of the risk of liquefaction during seismic loading (Adalier and Elgamal 2004; BenSalem et al. 2015; Tang et al. 2015). Several experimental studies have been developed to simulate the behavior of reinforced soil with stone columns against liquefaction such as triaxial tests, centrifuge and shaking table tests (Blewett and Woodward 2001; Adalier et al. 2003; Singh 2013). Several numerical analyses using the 2D finite difference code FLAC were performed to simulate the experimental tests.

T. Abdoun and S. Elfass (eds.), *Soil Dynamics and Soil-Structure Interaction for Resilient Infrastructure*, Sustainable Civil Infrastructures, DOI 10.1007/978-3-319-63543-9_6

The aim of this study is to evaluate the liquefaction susceptibility of silty sand reinforced by stone columns subjected to sinusoidal seismic motion in a shaking table test. The liquefaction potential of silty sand layer improved by stone columns was evaluated in terms of the maximum pore water pressure ratio r_u. In fact, soil liquefaction potential has often been defined when the pore water pressure, u, increased and became equal to the vertical effective stress ($r_u = 1.0$). For this purpose, a set of numerical analysis was carried out by using the 2D finite difference code FLAC. Shaking table test model was developed and validated with experimental results published by Ueng et al. (2010). The effect of: (i) the excitation frequency and amplitude and (ii) stone columns installation process on liquefaction potential of improved soil have been investigated in the present study.

2 Numerical Analyses

In the present research, two-dimensional (2D) model was developed to simulate liquefaction susceptibility of reinforced silty sand with stone columns subjected to seismic motion. Geotechnical properties of materials are obtained from data of oil storage tanks project where stone columns technique was used for liquefaction mitigation (Duzceer 2003). Plane strain numerical analysis was conducted using the 2D finite difference code FLAC. The Finn constitutive method was used to model pore pressure build up during seismic loading to assess the liquefaction potential. The model used for the numerical analysis was presented in Fig. 1 and is 30 m wide and 2 m high constituted by silty sand reinforced by vertical trench of stone material of 0.8 m thickness. As shown in Fig. 2, the used mesh grid is denser in the area of the improvement (i.e. close to the trench). The bottom boundary is fixed both in the horizontal and vertical directions and the lateral boundaries are just fixed in the horizontal direction. The seismic motion was applied at the base of the model and it was assumed to be a harmonic sinusoidal loading. The geotechnical properties of constitutive materials (i.e. silty sand and stone material) are provided in Table 1.

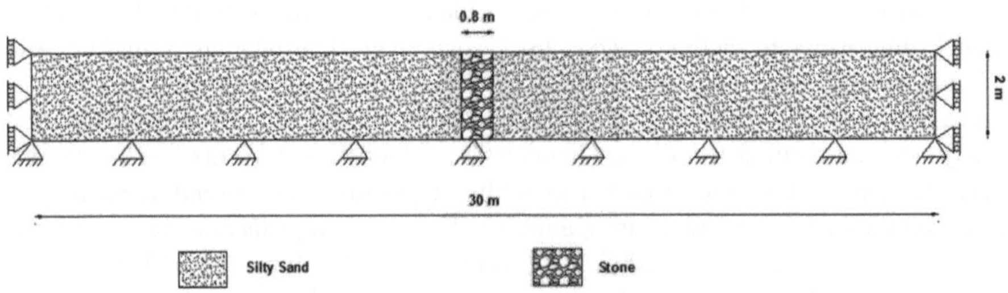

Fig. 1. Numerical model of reinforced soil

The numerical model was established based on the experimental data from shaking table tests published by Ueng et al. (2010). Series of seismic shaking tests performed in large biaxial laminar shear box were conducted by Ueng et al. (2010) on saturated clean

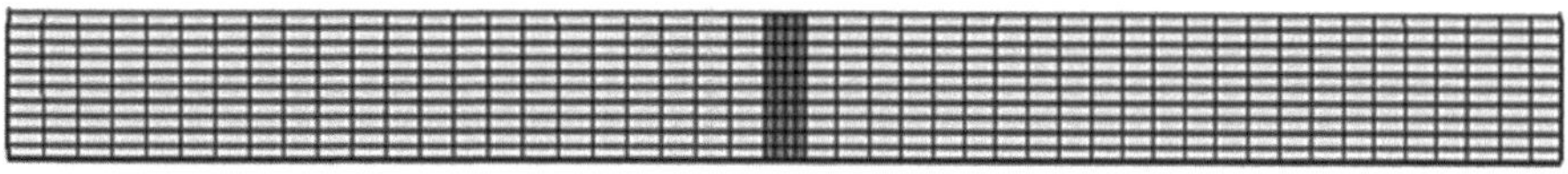

Fig. 2. Mesh grid

Table 1. Material properties

Soil	Parameter										
	E MPa	G MPa	K MPa	ϕ°	C kPa	k m/s	γ kN/m^3	D_r %	$N_{1,60}$	C_1	C_2
Silty sand	15	5.76	12.5	28	0.2	10^{-5}	20	41	8	0.64	0.62
Stone	30	11.5	25	40	0	10^{-3}	20	-	-	-	-

E:elastic modulus, G:shear modulus, K: bulk modulus, ϕ: friction angle, C: cohesion, k: permeability, γ:density,D_r:relative density, $N_{1,60}$:SPT test number, C1 and C2: constants used for Finn model.

Vietnam sand to evaluate liquefaction settlement under various seismic motions. The numerical model is verified for two different experimental tests E16 and F30. Details of input seismic motions of these experimental tests (Amplitude, frequency and duration) are given in Table 2 (Ueng et al. 2010). The variation of pore pressure at different depths of the soil layer obtained from the numerical simulation was compared with the measured pore water pressure during E16 and F30 tests in Figs. 3 and 4, respectively. It was observed that the variation of pore water pressure with depth obtained from the numerical model is close to experimental results.

Table 2. Characteristics of shaking table tests E16 and F30 (Ueng et al. 2010)

Test	E16	F30
Frequency (Hz)	2	4
Amplitude	0.075 g	0.075 g
Duration (s)	5	5

After model validation, numerical analyses were undertaken for reinforced silty sand with stone trench subjected to seismic motion by considering a peak acceleration $a_{max} = 0.05g$ and frequency of 4 Hz during 4 s. These simulations assumed a perfect drainage by the trench material.

Time histories of excess pore pressure ratio r_u recorded at depth of 0.3 m and located at 0.5 m, 1 m and 2 m from the border of stone trench are shown in Fig. 5. These time histories of excess pore pressure ratio r_u are compared with results for the unimproved soil. Figure 5 shows that during seismic motion the excess pore-pressure ratio builds up and reaches a peak value. For the unimproved soil, the peak value of excess pore pressure ratio r_u was about 0.91 indicating high liquefaction potential. Whereas, for improved soil, the peak values of r_u are smaller indicating that stone

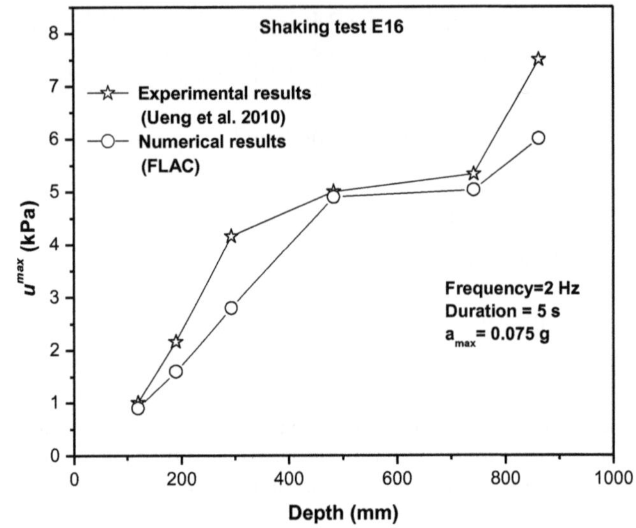

Fig. 3. Experimental and numerical excess pore water pressures (shaking test E16)

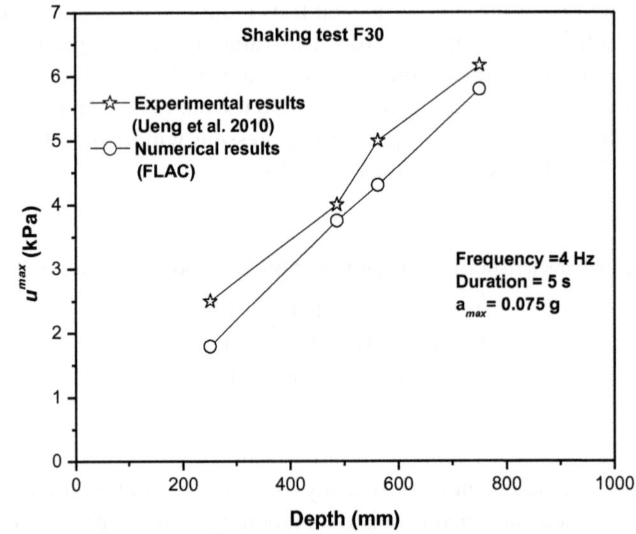

Fig. 4. Experimental and numerical excess pore water pressures (shaking test F30)

trench decreased the rate of generation of excess pore pressure. It can be also observed from Fig. 5 that the decrease in r_u values varied with horizontal distance from the stone trench. In fact, r_u values are smaller at the vicinity of stone trench and increase with horizontal distance. The peak values of r_u reached 0.38, 0.7 and 0.72 when recorded at 0.5 m, 1 m and 2 m distance from the stone trench, respectively. The rate of reduction

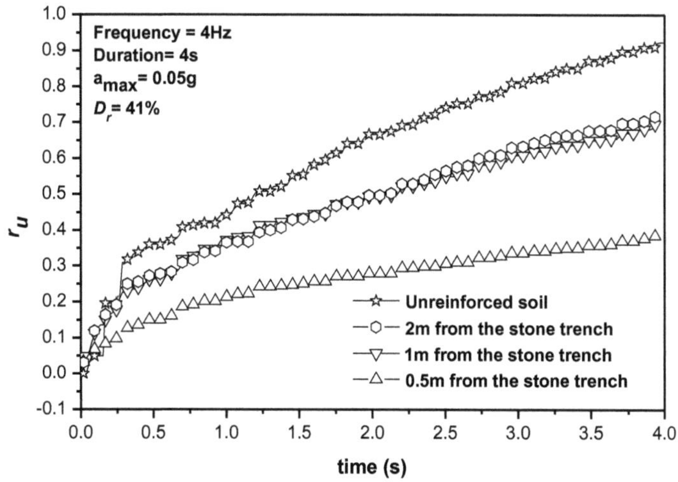

Fig. 5. Variation of r_u versus time at various distances from the stone trench

in the peak values of r_u after reinforcement $\left(\frac{r_u^{unimpr} - r_u^{impr}}{r_u^{unimpr}}\right)$ is 58.24%, 23.07% and 20.88% recorded at 0.5 m, 1 m and 2 m from the stone trench, respectively. A significant reduction in r_u values is observed near the stone trench and decreases with horizontal distance. These results confirm the drainage ability of stone material which is greater at the vicinity of reinforcing trench. Wide experimental researches from shaking table and centrifuge tests have confirmed these results (Iai 1988; Brennan and Madabhushi 2002).

3 Parametric Study

Parametric study was performed to investigate the effects of both shaking frequency, amplitude of excitation and stone columns installation process on liquefaction potential of the reinforced silty sand. Presented results were undertaken at the vicinity of the stone trench (i.e. 1 m away from the stone trench in depth of 0.3 m).

3.1 Effect of Frequency of Seismic Loading

Figure 6 shows the influence of seismic motion frequency on the variation of the excess pore pressure ratio r_u with time for reinforced silty sand with initial relative density D_r of 41%. Different frequencies of seismic motion are considered (6, 5, 4.5, 4, 3 and 2 Hz) at an acceleration of a_{max} = 0.05 g during 4 s. It can be observed that during the seismic loading, the excess pore pressure was generated and reached a peak value. In the case where the frequency of seismic motion is greater than 4.5 Hz, the peak values of r_u exceeded 1.0 and reached 1.4. In fact, when the excess pore pressure exceeds the initial effective stress means that the soil lost all the effective stress and was

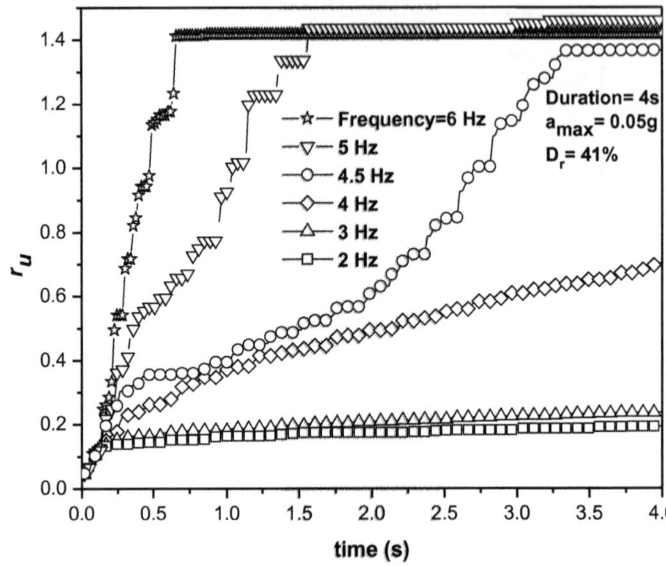

Fig. 6. Variation of r_u versus time for different frequency of seismic loading

in the unstable condition (Zen et al. 1987; Tsuchida et al. 2006). Therefore, it can be concluded that liquefaction occurs when the frequency of loading is greater than 4.5 Hz. However, with frequency of seismic loading less than 4.5 Hz (4, 3 and 2 Hz), the peak values of excess pore pressure ratio r_u were lower than 1.0. (0.7, 0.23 and 0.19 respectively). For those frequencies, liquefaction does not occur. As it can be seen in Fig. 6, when the frequency of seismic loading decreases, the rate of excess pore pressure generation decreases and then the liquefaction potential is reduced.

3.2 Effect of Maximum Amplitude of Seismic Loading

Different maximum amplitudes are selected for the seismic motion applied to the base of the model (a_{max} = 0.04 g, 0.05 g, 0.075 g, 0.1 g, 0.2 g and 0.3 g) at a frequency of 4 Hz during 4 s to evaluate its effect on liquefaction potential of reinforced silty sand with initial relative density D_r of 41%. Figure 7 shows the evolution of excess pore pressure ratio r_u with time for the considered maximum amplitudes. When the maximum amplitude of seismic motion a_{max} is greater than 0.075 g, the peak values of the excess pore pressure r_u exceeded 1.0. However, with a_{max} less than 0.075 g (a_{max} = 0.05 g and 0.04 g), the peak values of r_u were lower than 1.0. (0.7 and 0.48, respectively). By increasing the maximum amplitude of seismic motion a_{max}, the excess pore pressure ratio r_u increases enhancing liquefaction potential. It can be concluded that there is a significant effect of the maximum amplitude of seismic motion on the liquefaction potential of improved soil with stone columns.

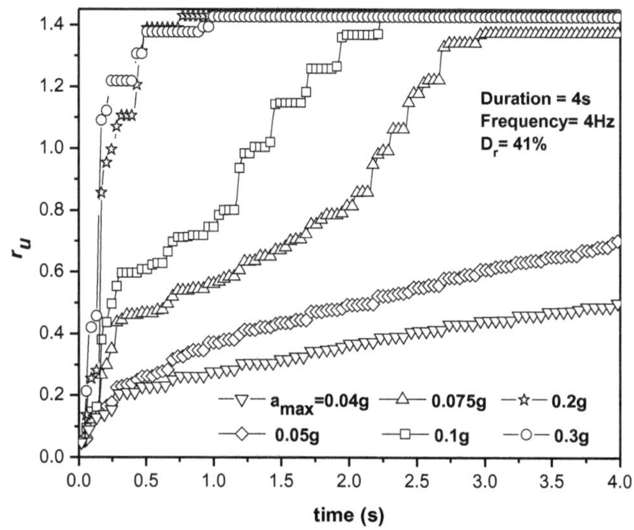

Fig. 7. Variation of r_u versus time for different a_{max}

As reported by many experimental investigations results, the effectiveness of using stone columns to mitigate liquefaction is largely affected by seismic motion characteristics (Sesov et al. 2004; Hamedi and Marandi 2011).

3.3 Effect of Stone Columns Installation

Vibro-stone columns installation process enhances densification of reinforced soil. In fact, the vibro-stone column installation process involves insertion of a vibratory probe. Densification of the soil around the stone column during installation occurs which is due to two phenomena: (i) during insertion of the vibratory probe, excess pore water pressure is induced in the soil surrounding the probe, and concurrent dissipation occurs leading to soil densification; and (ii) cavity expansions during repeated fillings of the cavity by stones and probe insertions cause significant excess pore water pressures, and concurrent consolidation and soil densification (Baez and Martin 1993; Debats et al. 2003; Shenthan 2005). Different relative densities are selected to investigate the effect of stone columns installation process on liquefaction potential. Initially, the relative density D_r of reinforced soil was equal to 41%. The rates of increase for D_r after stone column installation are considered about 46% and 95% as derived from in situ tests performed before and after reinforcement in many case histories (Blanchard and Clements 1993; Baez 1995; Weaver et al. 2004). Thus, different relative densities are considered; D_r = 60% and 80%. The applied seismic loading was at an acceleration of 0.05 g and a frequency of 0.4 Hz during 4 s. Figure 8 shows the increase in excess pore water pressure ratio r_u with time for different relative densities (41%, 60% and 80%). The case D_r = 41% (i.e. initial relative density before stone columns installation), the peak value of r_u was 0.7. Whereas, with D_r > 41%, the peak values of r_u

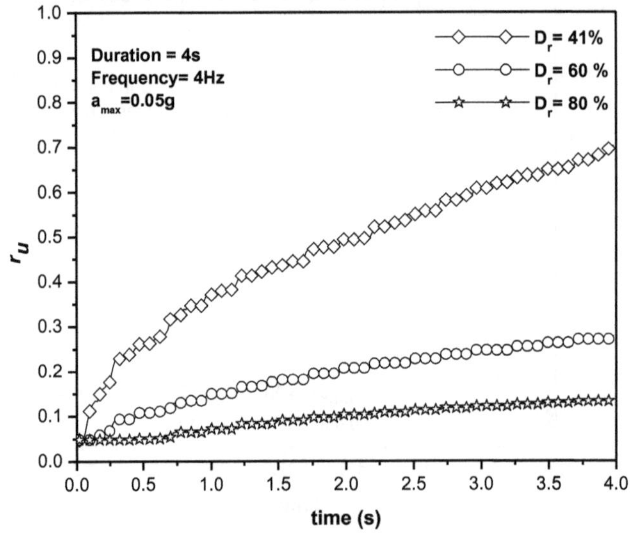

Fig. 8. Variation of r_u versus time for different D_r

were less than 0.7 (0.27 and 0.13, respectively). It can be concluded greater amount of reduction in peak values of r_u can be achieved with higher relative densities of reinforced soil. In fact, vibro-stone columns installation process increases the relative density of the initial soil which decreases excess pore pressure ratio r_u. Consequently, the liquefaction potential of reinforced soil is reduced. As reported by Andrus and Chung (1995) and Mitchell et al. (1995), the densified soil is more resistant to liquefaction and has performed well during earthquakes.

4 Conclusion

A numerical model was carried out using the 2D finite difference code FLAC to study the effectiveness of using stone columns to mitigate liquefaction. This model was developed to simulate the shaking table test and was validated with published experimental results from literature. Parametric analyses involving the effect of the applied excitation frequency and amplitude and the effect of the installation process of stone columns were performed.

Based on the results of this numerical study, it was found that the generation of excess pore pressure ratio decreases with stone columns reinforcement and liquefaction potential is significantly reduced. It was also shown that when the frequency and the amplitude of seismic motion decrease, the rate of excess pore pressure generation is decreased and liquefaction potential is reduced. The increase in relative density of reinforced soil due to vibro-stone columns installation process, leads to an improved liquefaction resistance.

References

Adalier, K., Elgamal, A.: Mitigation of liquefaction and associated ground deformations by stone columns. Eng. Geol. **72**(3–4), 275–291 (2004)

Adalier, K., Elgamal, A., Meneses, J.F., Baez, J.I.: Stone columns as liquefaction counter-measure in non-plastic silty soils. J. Soil Dyn. Earthq. Eng. **23**(7), 571–584 (2003)

Andrus, R., Chung, R.: Ground improvement techniques for liquefaction remediation near existing lifelines. Building and fire research laboratory, National Institute of Standards and Technology (1995). Report no. 5714. Gaithersburg

Baez, J.I.: A Design Model for the Reduction of Soil Liquefaction by Vibro-Stone Columns. Ph. D. Dissertation, University of Southern California, Los Angeles, CA (1995)

Baez, J.I., Martin, G.R.: Advances in the design of vibro systems for the improvement of liquefaction resistance. In: Proceedings of the Symposium of Ground Improvement, Vancouver Geotechnical Society, Vancouver, BC, pp. 1–16 (1993)

BenSalem, Z., Frikha, W., Bouassida, M.: Effect of granular-column installation on excess pore pressure variation during soil liquefaction. Int. J. Geomech. (2015). 04015046, 10.1061/(ASCE)GM.1943-5622.0000516

Blanchard, J.D., Clements, K.: Site improvements with stone columns in stratified silty soils. In: Third International Conference on Case Histories in Geotechnical Engineering, Missouri University of Science and Technology, St. Louis, Missouri, pp. 1027–1033 (1993)

Blewett, J., Woodward, P.K.: A triaxial system for vibro-replacement liquefaction mitigation studies. In: Proceedings of the ICE - Ground Improvement, pp. 75–83 (2001)

Brennan, A., Madabhushi, S.: Effectiveness of vertical drains in mitigation of liquefaction. Soil Dyn. Earthq. Eng. **22**(9–12), 1059–1065 (2002)

Debats, J.M., Guetif, Z., Bouassida, M.: Soft soil improvement due to vibro-compacted columns installation. In: Proceedings of the International Workshop on Geotechnics of Soft Soils - Theory and Practice, Noordwijkerhout, Netherlands, pp. 551–556 (2003)

Duzceer, R.: Ground improvement of oil storage tanks using stone columns. In: Proceedings of the 12th PanAmerican Conference in Soil Mechanics and Foundation Engineering, Essen: Verlag Glückauf GMBH, Cambridge, Massachusetts, pp. 1681–1686 (2003)

Hamedi, A., Marandi, S.M.: Laboratory studies on the effect of vertical gravel column drains on liquefaction potential. Int. J. Eng. Trans. B: Appl. Mater. Energy Res. Cent. **24**(3), 209–225 (2011)

Iai, S.: Large scale model test and analyses of gravel drains. Report of the Port and Harbour Research Institute Japan, 27(3) (1988)

Mitchell, J.K.: Mitigation of liquefaction potential of silty sands. In: Laier, J.E., Crapps, D.K., Hussein, M.H. (eds.) From Research to Practice in Geotechnical Engineering, Geotechnical special publication 180, ASCE, New Orleans, Louisiana, pp. 453–451 (2008)

Mitchell, J.K., Baxter, C.D.P., Munson, T.C.: Performance of improved ground during earthquakes. In: Soil Improvements for Earthquake Hazard Mitigation, Geotechnical Special Publication No. 49, ASCE, pp. 1–36 (1995)

Mitchell, J.K., Wentz, F.K.: Performance of improved ground during the Loma Prieta earthquake. Report No. EERC91/12, Earthquake Engineering Research Center, University of California, Berkeley, CA (1991)

Sesov, V., Harada, N., Towhata, I., Talaganov, K.: Experimental investigations on drain improvement method - reduction of liquefaction potential. In: 13th World Conference on Earthquake Engineering, Vancouver, BC, Canada (2004)

Shenthan, T.: Liquefaction mitigation in silty soils using composite stone column. Ph.D. Dissertation, University at Buffalo, Buffalo (2005)

Singh, H.P.: Effects of surcharge loads on liquefaction parameters of pond ash improved with stone-sand columns. Int. J. Civ. Eng. Technol. (IJCIET) **4**(4), 225–235 (2013)

Tang, L., Cong, S., Ling, X., Lu, J., Elgamal, A.: Numerical study on ground improvement for liquefaction mitigation using stone columns encased with geosynthetics. Geotext. Geomembr. **43**(2), 190–195 (2015)

Tsuchida, T., Yoshimuta, S., Asa-Umi, R.: Upward movement of fines in sand under one dimensional water pressure change. In: Proceedings of the Geomechanics and Geotechnics of Particulate Media, Japan, pp. 417–419 (2006)

Ueng, T.S., Wu, C.W., Cheng, H.W., Chen, C.H.: Settlements of saturated clean sand deposits in shaking table tests. Soil Dyn. Earthq. Eng. **30**(1–2), 50–60 (2010)

Weaver, T., Ashford, S., Rollins, K.: Performance and analysis of a laterally loaded pile in stone column improved ground. In: 13th World Conference on Earthquake Engineering, Vancouver, BC, Canada (2004)

Zen, K., Yamazaki, H., Watanabe, A.: Wave-induced liquefaction and densification in seabeds. Report of Port and Harbour Research Institute, vol. 26(4), pp. 125–180 (1987)

Uplift Resistance of Offshore Pipelines Subject to Upheaval Buckling

Sahar Ismail(✉), Shadi Najjar, Salah Sadek, and Mounir Mabsout

Department of Civil and Environmental Engineering, University of Beirut,
P.O. Box 11-0236 Riad El-Solh, Beirut 1107-2020, Lebanon
{sai13, sn06, salah, mounir}@aub.edu.lb

Abstract. Offshore pipelines that transport oil and gas in different areas in the world are often buried in trenches to provide stability and protection against upheaval buckling. Hydrocarbons in the pipeline are transported at high temperatures and pressures to facilitate the oil flow and prevent its solidification. However, this mode of transport causes an increase in the axial compressive forces inside the pipeline which may lead to upward buckling in the direction of the least soil resistance. Upheaval buckling can result in pipeline failure causing severe environmental and economic losses. In this paper, a 3D parametric study of upheaval buckling of pipelines buried in medium dense sand with fines is performed using the finite element software Abaqus©. The effects of pipeline diameter, embedment depth ratio and diameter to wall thickness ratio on the soil resistance against uplift are investigated for pipeline pullout cases simulating plane strain conditions. The results are compared with the available analytical/empirical solutions for the uplift resistance. The results show that the uplift resistance depends in part on the pipeline diameter and embedment depth ratio. The available design methods which assume mobilized soil blocks above the pipeline with inclined slip surfaces better capture the behavior observed in the numerical models, as compared to those methods in which vertical slip surfaces are considered.

1 Introduction

Offshore pipelines transporting oil and gas placed on the seabed or buried in a trench are susceptible to global buckling. Transportation of oil and gas is typically conducted at high temperature and pressure to facilitate flow and prevent hydrocarbon solidification. However, this mode of transportation increases the axial forces inside the pipeline walls and may force the pipeline to buckle in the lateral or upward direction depending on whether the pipelines were laid on, or buried in, the seabed. Buckling can result in pipeline failure that can, not only cause severe economical losses, but also severe environmental damages to the seabed, shoreline, etc. Some examples of recent offshore oil leaks are those that occurred in Deepwater Horizon in the Gulf of Mexico (April 2010) and in Bohai Gulf in China (June 2011).

Studies on upheaval buckling analysis of subsea pipelines date back to 1974 (Palmer and Baldry 1974). However, it was not until a series of pipelines incidents that occurred in the 1980s in the North Sea and the increasing interest in the exploitation of

T. Abdoun and S. Elfass (eds.), *Soil Dynamics and Soil-Structure Interaction for Resilient Infrastructure*, Sustainable Civil Infrastructures, DOI 10.1007/978-3-319-63543-9_7

offshore gas and oil reserves worldwide that research about this topic increased. Researchers tackled upheaval buckling analyses through analytical approaches that are based on: (1) the "Vertical Slip Models" by Schaminee et al. (1990), Palmer et al. (1990), and Bransby et al. (2002), (2) the "Sliding Block with Inclined Failure Surfaces Models" by White et al. (2001), Ng and Springman (1994), and Vermeer and Sutjiadi (1985), and (3) codes and standards such as DNV (2007) and ASCE (1984, 2001) which present design procedures for pipelines. Examples of typical Vertical Slip Models and Inclined Failure Surface Models are presented in Fig. 1.

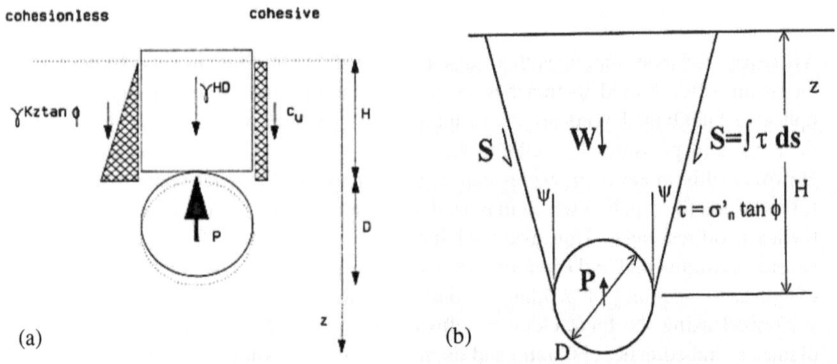

Fig. 1. (a) The vertical slip model (after Schaminee et al. (1990)) and (b) Sliding block mechanism with inclined shear planes (after White et al. 2001)

Upheaval buckling of buried offshore pipelines was also studied experimentally by Schaminee et al. (1990), Ng and Springman (1994), Bransby et al. (2001), Chin et al. (2006), Schupp et al. (2006), Stone (2006), Byrne et al. (2008, 2013), Cheuk et al. (2008), Bransby and Ireland (2009), Gao et al. (2011), Wang et al. (2012) and Robert and Thusyanthan (2014) and numerically by Yimsiri et al. (2004), Berghe et al. (2005), Newson and Deljoui (2006), Liu et al. (2008, 2013, 2014, 2015), Sun et al. (2011), Gao et al. (2011), Robert and Thusyathan (2014) and Zeng et al. (2014).

The majority of published studies that target the uplift resistance of soils in applications involving upheaval buckling of pipelines focus on pipelines that are embedded in idealized soil types (either sand or clay). For cases involving idealized sands, the shear strength of the sand is usually designated with an internal friction angle only. In reality, offshore pipelines are rarely embedded in clean sands. The offshore geologic settings that have to be crossed by pipelines could include sands that have different percentages of fines. The uplift resistance that is relevant to the design of offshore pipelines in silty sands has been rarely investigated, and the available idealized empirical models that are based solely on pure sands may not adequately model the case with silty sands that may exhibit a non-zero cohesive intercept upon uplift. A non-zero cohesive intercept may also be necessary for modeling the Mohr-Coulomb failure envelops of sands and silty sands in the low confinement pressure range that is relevant to problems involving pipelines that are buried to relatively small height to diameter ratios.

The objective of this paper is to fill a gap in the literature on the response of pipelines that are subjected to upheaval buckling in medium dense sands with fines. This is accomplished using a numerical finite element analysis in which the pipeline diameter, soil embedment depth, fines content (as reflected in the cohesive intercept) and diameter to wall thickness ratio are varied within a parametric study. The analysis is based on a 3D displacement-controlled finite element model that is solved within the commercial software Abaqus© by adopting elements which allow for the representation of the soil medium in its complex solid/pore fluid interaction. The results of the FE analyses are used to investigate the effect of the different parameters on the uplift response and to investigate the effectiveness of available empirical models in predicting the uplift resistance of pipelines in silty sands.

2 Finite Element Analysis

2.1 Finite Element Model

The finite element software, Abaqus© 6.13-4, was used to analyze the upheaval buckling behavior of buried offshore pipelines in medium dense sand with fines. The geometric model consisted of the pipeline and the subsoil. The diameter of the pipeline (D), the embedment depth ratio (H/D) and the diameter to wall thickness ratio (D/WT) constituted the main geometric parameters that were varied in the analyses. Only a quarter of the problem was simulated due to symmetry as shown in Fig. 2. The length of the pipeline was taken to be 100 m. The pipeline was modeled using eight node linear brick, reduced-integration, hourglass control C3D8R elements with around 3500 elements, while the subsoil was modeled using eight-node brick, trilinear displacement, trilinear pore pressure, reduced integration, hourglass control C3D8RP elements with around 90000 elements. The simulations were performed using large strain formulations while incorporating geometric nonlinearity using the NLGEOM option in Abaqus©. For the boundary conditions, the bottom soil boundary was assumed to be fixed while the movement of the soil wall boundaries and pipeline ends were restrained axially (Fig. 2).

The interaction between the pipe and the soil was modeled using an interface Mohr-Coulomb failure model (Yimisri et al. 2004) with an interface friction angle that is taken to be 0.5 of the peak friction angle of the surrounding soil. To account for possible strain-softening in the stress-strain response, the sand was modeled using a Modified Mohr-Coulomb model (MMC) that has the advantage of capturing the strain-softening behavior. A non-zero cohesive intercept was included in the constitutive model to reflect the presence of fines in the sandy soil. The softening behavior is generally modeled by a reduction of the mobilized friction and dilation angles with the increase in plastic deviatoric shear strain (Robert and Soga 2010; Robert and Thusyanthan 2014).

In the FE analysis, different pipeline diameters (D = 0.3, 0.6 and 0.8 m), different embedment depth ratios (H/D = 1, 2 and 3), two fines contents (reflected in c = 2 and 10 kPa) and three diameter to wall thickness ratios (D/WT = 25, 30 and 40) were adopted. The problem was built in 3D to allow for realistic vertical pressures on the

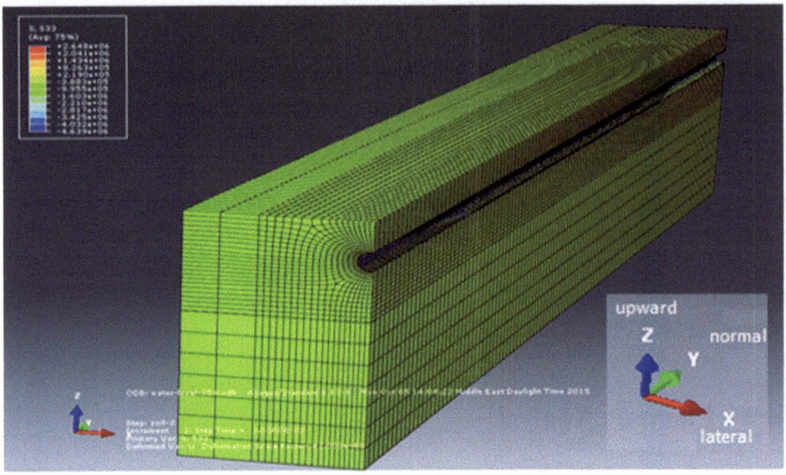

Fig. 2. Geometry and model mesh distribution

pipeline due to the application of the self-weight of the soil and the water above the pipeline. A depth of water of 100 m was adopted since it was considered to be representative of relatively shallow coastal waters in Lebanon and other areas in the world with operating offshore pipelines (ex. Bohai Gulf in China). The 3D model also allows for attaining a realistic settlement profile of the pipeline in the vertical direction due to its self-weight and the weight of overlying soil and water.

Following the application of body forces, the pipeline was uplifted along its whole length to simulate plane strain conditions and the force-displacement response was investigated. The resulting total soil resistance F is in units of force normalized by the length of the pipe (kN/m). For the sake of presentation, the total uplift resistance was further normalized with respect to the effective unit weight of the soil (Υ'), depth to the middle of the pipeline ($H_c = H + D/2$, with H being the depth to the crown of the pipeline) and the pipeline diameter (D). The resulting normalized upward force is $F/(\Upsilon' H_c D)$. The pipeline movement (disp) required to mobilize the maximum reaction force was also determined and normalized by the depth to the middle of the pipeline H_c. Tables 1 summarizes the soil and pipeline parameters and conditions used in the finite element study.

Table 1. Soil and pipeline parameters used in the finite element analyses

Soil properties									
E(MPa)	ν	Υ_{sat}(kN/m3)	Υ_d(kN/m3)	Gs	e	φ_{peak}°	φ_{res}°	ψ_{peak}°	ψ_{res}°
25	0.3	19	15	2.6	0.7	40	33	8.75	0.1
Soil dimensions				Pipe/Steel properties					
Length (m)	Height (m)	Width (m)		E (GPa)	ν	ρ (kg/m^3)	σ_y (MPa)		
100	10	10		206	0.3	7850	448		

2.2 Numerical Results and Analyses

2.2.1 Pipeline Diameter

The variation of the uplift resistance with normalized pipeline displacement is presented in Fig. 3 for cases involving pipelines with diameters of 0.3 m, 0.6 m, and 0.8 m. The cases pertain to an H/D of 2.0, a pipeline diameter to wall thickness ratio of 30 and to the case where the soil has a cohesion of 10 kPa. The curves on Fig. 3 indicate that the diameter of the pipeline has a significant effect on the uplift force versus displacement relationship. As the diameter of the pipeline increases, the initial slope of the force-displacement response becomes steeper and the maximum uplift force increases. The higher initial stiffness and the higher uplift resistance for cases involving larger diameters are generally attributed to the higher level of confinement and the greater volume of soil that is involved in the uplift resistance.

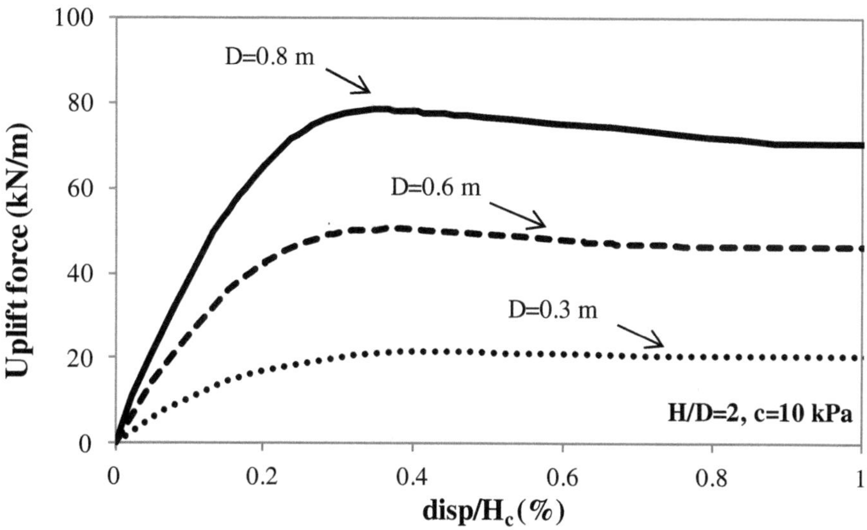

Fig. 3. Force-displacement curves for different pipeline diameters at H/D = 2 and c = 10 kPa

The load-displacement curves presented in Fig. 3 exhibit a strain-softening behavior that is expected for the modified Mohr-Coulomb model that is adopted in the FE analysis. The maximum uplift resistance increases from about 22 kN/m to about 80 kN/m as the pipeline diameter increases from 30 cm to 80 cm, respectively. The ultimate uplift resistance is mobilized at normalized displacements (disp/H_c) that vary in the narrow range of 0.3% to 0.4%, with the lower values being associated with the larger pipeline diameters. The FE results are in line with those reported by Liu et al. (2013).

2.2.2 Embedment Depth Ratio (H/D)

To investigate the impact of the embedment depth ratio (H/D) on the uplift force versus normalized displacement relationship, results pertaining to a pipeline diameter of 0.8 m, a diameter to wall thickness ratio of 30 and a soil with a cohesion of 10 kPa are presented in Fig. 4 for cases with H/D of 1, 2, and 3. As the embedment depth increases, results indicate that the soil resistance (block weight and shearing resistance along uplifted soil block boundaries) increases resulting in an increase in the uplift resistance for any normalized pipeline displacement. For comparison, the uplift force F increases from around 48 kN/m to around 113 kN/m as H/D increases from 1 to 3. The increase in uplift resistance is associated with an increase in the initial slope (stiffness) of the force-displacement relationship. These results are important since they indicate that increasing the embedment depth of a pipeline has a significant impact on the uplift resistance, rendering the H/D ratio as one of the most important design parameters for problems where the upheaval buckling of the pipeline is a source of risk in the design. The effect of pipeline embedment depth was explored by a number of researchers such as Schaminee et al. (1990), Ng and Springman (1994), Robert and Thusyathan (2014) and Liu et al. (2015). These studies showed that the uplift resistance increases systematically with the embedment depth as observed in the FE results on Fig. 4.

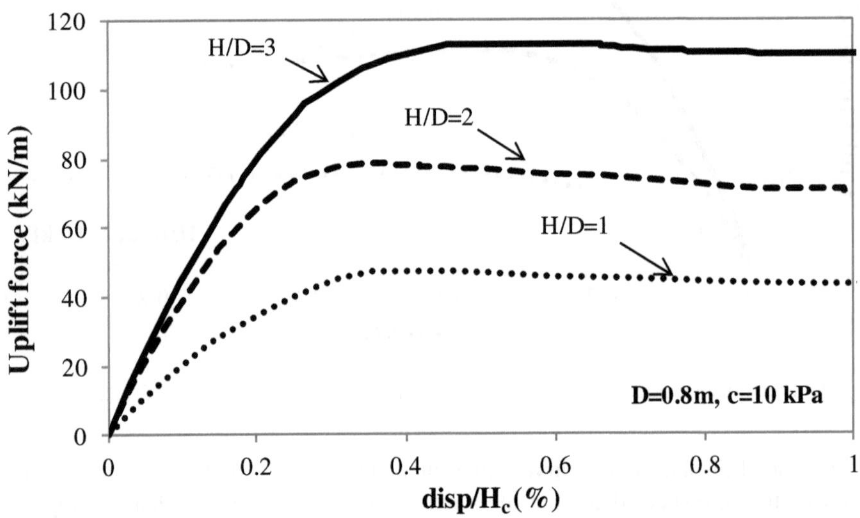

Fig. 4. Force-displacement curves for different embedment depth ratios for D = 0.8 m and c = 10 kPa

2.2.3 Fines Content

Offshore seabed deposits vary in composition and relative density. It is expected that offshore pipelines could be embedded in sands that include a certain percentage of fines. To model the presence of fines, identical finite element analyses were conducted for cases involving a cohesive intercept of 2 kPa and 10 kPa, respectively. For illustration, force-displacement curves pertaining to the case with a diameter of 0.8 m, a diameter to wall thickness ratio of 30 and an H/D of 2.0 are presented in Fig. 5 for cohesive intercept values of 2 and 10 kPa. Interestingly, results on Fig. 5 indicate that a small increase in the cohesive intercept could result in significant improvement in the ultimate uplift force, without affecting the initial stiffness. The maximum uplift force is found to increase from around 50 kPa to around 80 kPa, for the case were "c" increases from 2 kPa to 10 kPa. In addition, the normalized mobilization distance also increases as the soil cohesion increases.

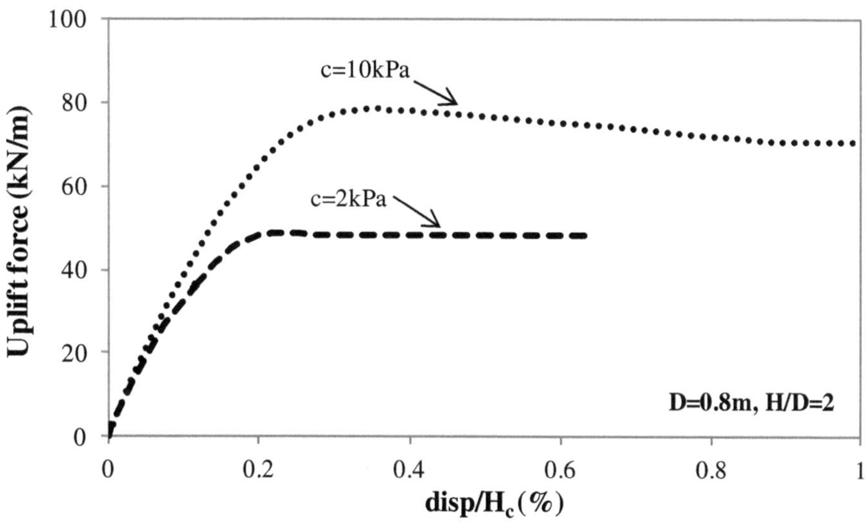

Fig. 5. Force-displacement curves for different soil cohesion values for D = 0.8 m and H/D = 2

The significant effect of the cohesive intercept on the maximum uplift force could be attributed to the fact that the uplift resistance is generally associated with small embedment depths and small confining levels (1 to 3 m of embedment). As a result, the frictional component of the resistance could become small (due to the lower stress levels) relative to any cohesive component of strength that could be introduced into the problem (c = 10 kPa versus 2 kPa). According to Schaminee and his co-workers, the cohesion component increases the soil strength by $2.H_c.c_u$, as is demonstrated in the analytical solution part included in the next section, and causes the pipe to displace more to reach the peak uplift resistance. As the pipe diameter and embedment depth ratio increase, the effect of the fines content (increasing cohesion) on the uplift soil resistance is expected to decrease relative to cases that involve smaller H/D ratios.

2.2.4 Diameter to Wall Thickness Ratio

In practice, the wall thickness design criteria for a high pressure high temperature pipeline depends on limiting the hoop stress due to internal pressure and hydrostatic collapse and buckle propagation due to external pressure. The minimum wall thickness should be greater than the design wall thickness from the above criteria. DNV (2007) recommends a diameter to wall thickness ratio (D/WT) between 15 and 45 with a minimum wall thickness of 12 mm. In this study, typical diameter to wall thickness ratios of 25, 30 and 40 were examined for the case of 0.8 m diameter pipeline embedded at H/D = 2 and for the case of c = 10 kPa. As presented in Fig. 6, the diameter to wall thickness ratio, taken within the DNV recommended range, slightly affects the uplift resistance. Thus, within the range of values tested, D/WT has a minimal influence on the pipe uplift resistance.

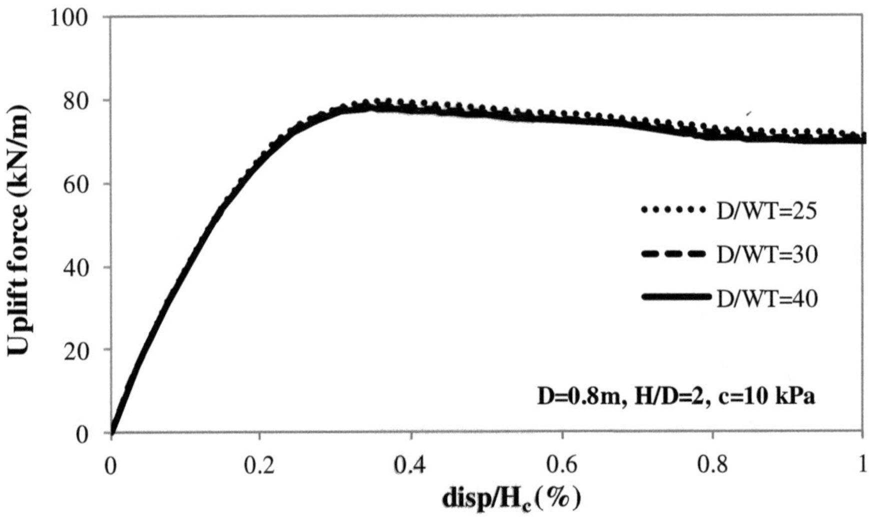

Fig. 6. Force-displacement curves for different diameter to wall thickness ratios for D = 0.8 m, H/D = 2 and c = 10 kPa

2.2.5 Normalization of Uplift Resistance

To combine the effects of the soil unit weight, pipeline embedment depth, and pipeline diameter on the FE-computed maximum uplift force, F was normalized by the above mentioned parameters and the results were plotted on Fig. 7 for all the cases analyzed in this study. Except for the specific case involving the smallest pipeline diameter of 0.3 m and the highest cohesive intercept of 10 kPa, the results show that the variation of the normalized upward force $F/(\Upsilon' H_c D)$ with normalized pipeline displacement is independent of the pipeline embedment depth. The variation of $F/(\Upsilon' H_c D)$ with displacement is observed to be highly dependent on the cohesive intercept, confirming that the response of pipelines under upheaval buckling is likely to be sensitive to the presence of fines within a sandy overburden soil. For the specific case involving a small diameter pipeline with a D = 0.3 m, the lack of normalization in the results for different H/D could be directly linked with the presence of a large cohesive intercept, which will play a dominant role in controlling the uplift resistance particularly at the shallow embedment depths (between 0.3 m and 0.9 m) that are associated with the H/D values adopted.

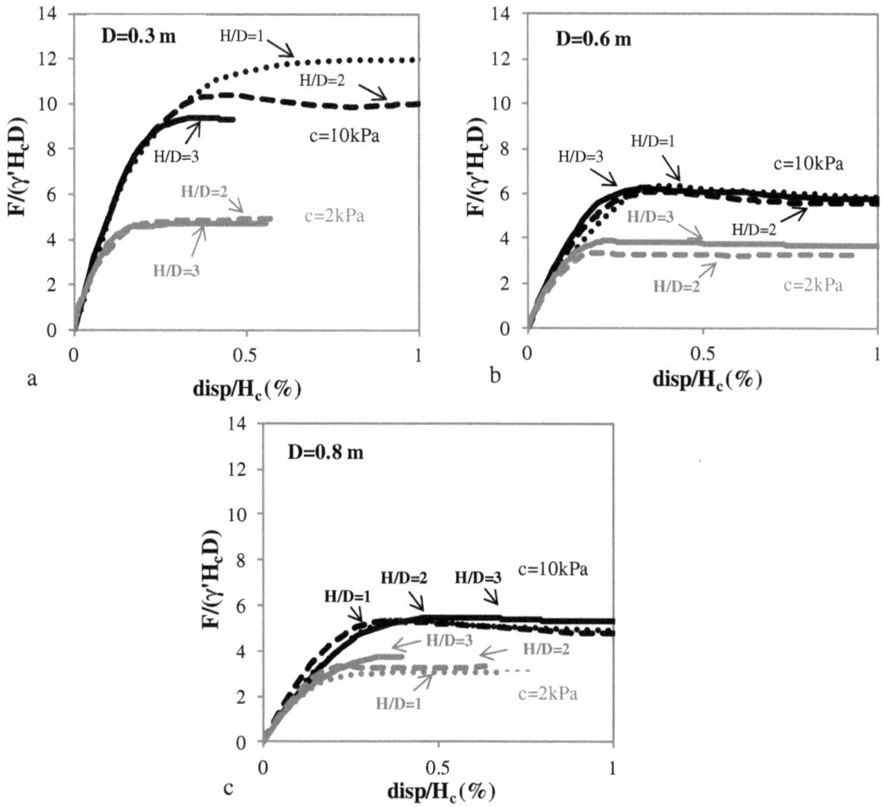

Fig. 7. Variation of normalized uplift resistance force with the normalized pipeline displacement for (a) D = 0.3 m, (b) D = 0.6 m and (c) D = 0.8 m

3 Analytical Solutions of Upheaval Buckling of Buried Offshore Pipelines

The key factor associated with the resistance to uplift of buried pipelines is the uplift resistance force, *F*, provided by the embedment soil, expressed in units of F/L (e.g. kN/m). Several researchers presented empirical formulae to obtain the maximum uplift resistance force. These formulae vary in reference to the models and assumptions used to derive them. The soil uplift resistance against upheaval can be derived from three sets of models. The first set of models (Vertical Slip Models) assume that the uplift soil resistance corresponds to the dead weight of the soil above the pipe along with the magnitude of friction and cohesion being mobilized along the failure surfaces for cohesionless and cohesive soils, respectively. The second set of models (Sliding Block with Inclined Failure Surfaces) considers that the shearing plane is inclined at an angle equal to the dilation angle. The inclined block results in additional soil weight, changes in the length of the failure surface, and changes in the normal stresses on the shear plane (Fig. 1). A third model (DNV, DET NORSKE VERITAS) describes the uplift soil resistance using the parameter R_{max} (DNV 2007), which takes into consideration the weight of the soil column above the pipeline combined with a cohesive resistance component for the global shallow shear failure mode. The mathematical formulation that describes the different predictive models are presented in Eqs. 1 to 7 where the maximum uplift resistance provided by the soil against upheaval buckling is given as:

$$F = \Upsilon' HD + H^2\gamma' K\tan\Phi \quad \text{(Sands - Shaminee etal. (1990))} \tag{1}$$

$$F = \Upsilon' HD + 2Hc_u \quad \text{(Clays - Shaminee etal. (1990))} \tag{2}$$

$$F = \Upsilon' HD[1 + 0.1D/H + K_0\tan\Phi(H/D)(1 + D/(2H))^2] \quad \text{(Sands - Bransby etal. (2002))} \tag{3}$$

$$F = \Upsilon' HD[1 + 0.1D/H + 2c_u/(\Upsilon' H)(H/D + 0.5)] \quad \text{(Clays - Bransby (2002))} \tag{4}$$

$$F = \Upsilon' HD + \Upsilon' H^2\tan\psi + \Upsilon' H^2(\tan\Phi_{peak} - \tan\Psi)[(1+K_0)/2 - (1 - K_0)(\cos 2\Psi)/2 \quad \text{(White etal. 2001)} \tag{5}$$

$$F = (\Upsilon' H_c D)(1 + fH_c/D)\ \text{with}\ f = K\tan\Phi \quad \text{(Sands - DNV (2007) using}\ R_{max}) \tag{6}$$

$$F = \Upsilon' HD + \Upsilon' D^2(0.5 - \pi/2) + 2c_u(H + D/2) \quad \text{(Clays - DNV (2007) using}\ R_{max}) \tag{7}$$

Where $K = k_o = 1 - \sin\Phi$ for loose sands and $K = K_p$ for medium dense and dense sands, and $f = \tan\Phi/\text{sqrt}((1 + \tan2\Phi) - \tan\varphi\text{sqrt}(1 + r)^2)^2$ with $r = -1$.

The results of the FE model were compared to the analytical solutions resulting from the application of Eqs. 1 to 7 to predict the maximum uplift force for the different cases analyzed. To allow for a realistic comparison between analytical and numerical (FE) results for the cases involving soils with a friction angle and a non-zero cohesive intercept, the predictions of the analytical models were formulated by adding the contributions of the frictional and cohesive components of strength. This was implemented by "merging" the predictions from "sand" and "clay" equations for each model to produce a "combined hybrid solution". For the case of the White et al. (2001) model which includes only a solution for cohesionless soils, the contribution of the cohesive intercept to the total uplift resistance was estimated to be equal to $2.H_c.c_u$ as is the convention.

The analytically predicted uplift resistances were calculated and plotted on Fig. 8 versus the numerically quantified resistances for the different analytical models considered. Results on Fig. 8 lead to the following main observations:

1. The Vertical Slip Model that is presented in Shaminee et al. (1990) seems to under predict the FE uplift resistance for all the cases considered, irrespective of the cohesive intercept and the embedment ratio. The average ratio of the FE resistance to the analytical resistance is around 1.92 for the case with a c = 2 kPa and 1.6 for the case with a c = 10 kPa.
2. The performance of the Vertical Slip Model that is presented in Bransby et al. (2002) is slightly better with an average ratio of the FE resistance to the analytical resistance of around 1.54 for the case with a c = 2 kPa and 1.23 for the case with a c = 10 kPa.

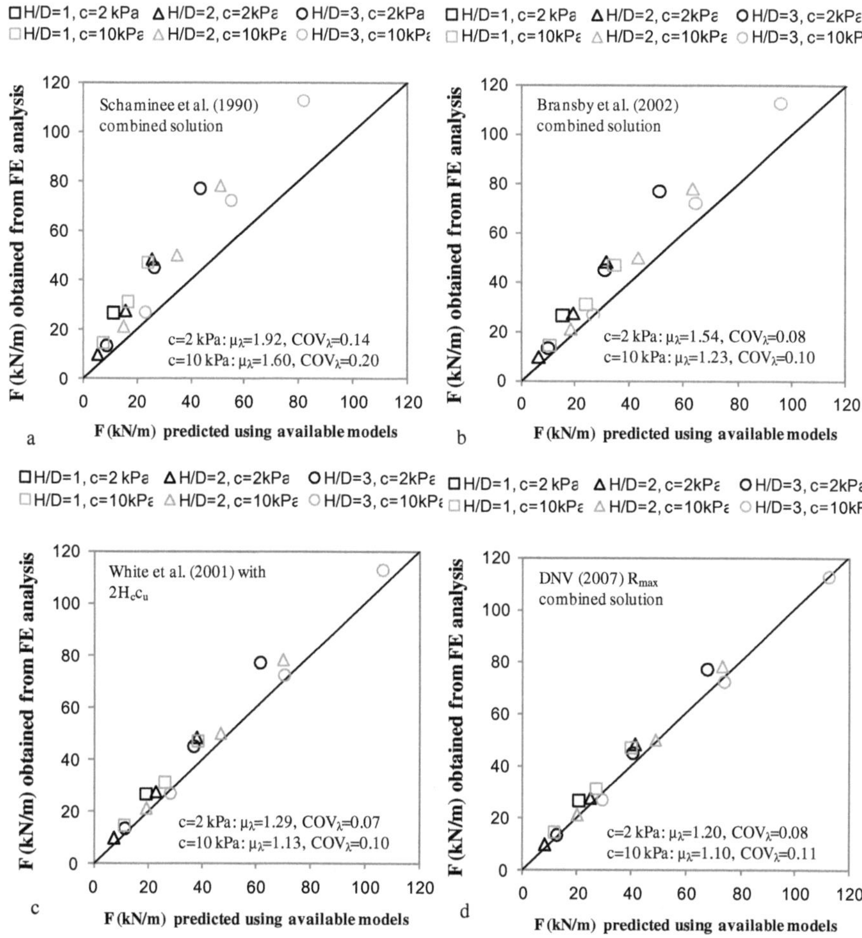

Fig. 8. Maximum uplift force obtained from FE analysis versus maximum uplift force predicted using (a) Schaminee et al. (1990), (b) Bransby et al. (2002), (c) White et al. (2001) and DNV (2007) using R_{max} combined solutions

3. For the Inclined Slip Model presented in White et al. (2001), the combined solution with $2.H_c.c_u$ exhibits predictions that are closer to the FE results with average bias factors that are equal to 1.29 for the c = 2 kPa case and 1.13 for the c = 10 kPa case.
4. Finally, results from the DNV (2007) model using the R_{max} combined solution results in the closest predictions to the FE results with average bias factors of about to 1.20 for c = 2 kPa and 1.10 for c = 10 kPa.

4 Conclusions

Buried offshore pipelines transporting oil and gas at high temperature and pressure are subjected to upheaval buckling. To study the effects of pipeline diameter, soil embedment depth, fines content and diameter to wall thickness ratio, 3D displacement-controlled finite element models were performed using the commercial software Abaqus©. Pipelines of 0.3, 0.6 and 0.8 m diameters with diameter to wall thickness ratios of 25, 30 and 40 were buried in medium dense sand with fines (2 and 10 kPa cohesion values) at embedment depth ratios H/D of 1, 2 and 3. The pipelines were pulled in plane strain conditions to obtain force-displacement responses. The FE results were compared to the available analytical solutions.

The results showed that the uplift soil resistance and mobilization distance increase with the pipeline diameter, embedment soil depth as well as with the fines content due to the increase in the effect of soil resistance (block weight and shearing resistance along uplifted soil block boundaries) and the added contribution of cohesion to the total capacity. The diameter to wall thickness ratio, taken within DNV (2007) recommended range, slightly affects the uplift resistance. For the combined analytical solutions, the Inclined Slip Model that is presented in White et al. (2001) and modified to include the contribution of cohesion through $2.H_c.c_u$ and the DNV (2007) model using R_{max} resulted in predictions that are the closest to the FE results, outperforming the traditional Vertical Slip Models.

Notation

The following symbols are used in this paper:

F	= Peak uplift resistance per unit length (kN/m)
R_{max}	= Maximum uplift resistance per unit length (kN/m)
Υ_d	= Dry unit weight (kN/m^3)
Υ_{sat}	= Saturated unit weight(kN/m^3)
Υ'	= Effective unit weight (kN/m^3)
c_u	= Undrained shear strength (kPa)
c	= Soil cohesion (kPa)
E	= Young Modulus (Pa)
ν	= Poisson's ratio
D	= Outside pipe diameter (mm)
WT	= Pipeline wall thickness
H	= Soil cover height; depth to the top of the pipe (m)
H/D	= Embedment depth ratio
K	= Coefficient of lateral earth pressure
Ψ	= Dilation angle (°)
Φ	= Internal friction angle (°)
Φ_{peak}	= Peak friction angle (°)
Φ_{crit}	= Critical friction angle (°)
Φ'_{max}	= Maximum friction angle (°)

Ψ_{peak} = Peak dilation angle (°)
FE = Finite Element
μ = Interface friction coefficient

References

Abaqus Analysis User's Manual (version 6.13-4). Abaqus, Inc. (2016)

ASCE: Guidelines for the seismic design of oil and gas pipeline systems. Committee on Gas and Liquid Fuel Lifelines of the ASCE Technical Council on Lifeline Earthquake Engineering, Reston, VA (1984)

ASCE: Guidelines for the design of buried steel pipe. American Lifelines Alliance. ASCE, New York (2001)

Bransby, M.F., et al.: Numerical and centrifuge modeling of the upheaval resistance of buried pipelines. In: Proceeding of the 20th International Conference on Offshore Mechanics and Arctic Engineering, Rio De Janeiro, Brazil, OMAE2001/PIPE-4118 (2001)

Bransby, M.F., et al.: The upheaval capacity of pipelines in jetted clay backfill. Int. J. Offshore Polar **12**(4), 280–287 (2002)

Bransby, M.F., Ireland, J.: Rate effects during pipeline upheaval buckling in sand. In: Proceeding of the Institution of Civil Engineers (2009). doi:10.1680/geng.2009.162.5.247

Byrne, B.W., et al. Experimental modelling of the unburial behaviour of pipelines. In: Proceeding of Offshore Technology Conference, Houston, TX, Paper OTC 19573 (2008). doi:10.4043/19573-MS

Byrne, B.W., et al.: Uplift of shallowly buried pipe sections in saturated very loose sand. Geotechnique (2013). doi:10.1680/geot.11.P.016A

Cheuk, C.Y., et al.: Uplift mechanism of pipes buried in sand. J. Geotech. Environ. Eng. (2008). ASCE. doi:10.1061/(ASCE)1090-0241(2008)134:2(154)

Chin, E.L., et al.: Uplift resistance of pipelines buried in cohesionless soil. In: Ng, C.W.W., Wang, Y.H., Zhang, L.M. (eds.) Proceeding of the 6th International Conference on Physical Modelling in Geotechnics, vol. 1, pp. 723–728. Taylor and Francis Group, London (2006)

Det Norske Veritas (DNV): Global buckling of submarine pipelines-structural design due to high temperature high pressure. RP-F110, Oslo, Norway (2007)

Gao, X., et al.: Model test based soil spring model and application in pipeline thermal buckling analysis. China Ocean Eng. (2011). Springer. doi:10.1007/s13344-011-0041-6

Liu, R., et al.: Numerical studies on global buckling of subsea pipelines. J. Ocean Eng. (2014). Elsevier. doi:10.1016/j.oceaneng.2013.12.018

Liu, R., et al.: Laboratory tests and thermal buckling analysis for pipes buried in Bohai soft clay. J. Mar. Struct. (2015). Elsevier. doi:10.1016/j.marstruc.2015.05.001

Liu, J.X., et al.: Study examines causes of upheaval buckling in shallow subsea PIP lines. J. Oil Gas **106**(6), 53–57 (2008)

Liu, R., et al.: Finite element analysis on thermal upheaval buckling of submarine burial pipelines with initial imperfections. J. Cent. South Univ. Press (2013). Springer, Heidelberg. doi:10.1007/s11771-013-1481-3

Newson, T.A., Deljoui, P.: Finite element modeling of upheaval buckling of buried offshore pipelines in clayey soils. In: Proceedings of the Geoshangahai 2006 Conference on Soil and Rock Behavior and Modelling. ASCE (2006). doi:10.1061/40862(194)47

Ng, C.W.W., Springman, S.M.: Uplift resistance of buried pipelines in granular materials. In: Leung, C.F., Lee, F.H., Tan, T.S. (eds.) Centrifuge 1994, pp. 753–758 (1994)

Palmer, A.C., Baldry, J.A.S.: Lateral buckling of axially compressed pipelines. J. Pet. Technol. (1974). Springer. doi:10.2118/4815-PA

Palmer, A.C., et al.: Design of submarine pipelines against upheaval buckling. In: Proceeding of the 22nd Offshore Technology Conference, Richardson, TX, pp. 540–550 (1990). doi:10.4043/6335-MS

Robert, D.J., Thusyanthan, N.I.: Numerical and experimental study of uplift mobilization of buried pipelines in sands. J. Pipeline Syst. Eng. Pract. (2014). ASCE. doi:10.1061/(ASCE)PS.1949-1204.0000179

Robert, D.J., Soga, K.: Soil pipeline interaction in unsaturated soils. In: New Trends in the Mechanics of Geomaterials. Wiley, Lausanne (2010). Chap. 13

Schaminee, P., et al.: Soil response for pipeline upheaval buckling analyses: full-scale laboratory tests and modeling. In: Proceedings of the 22nd Annual Offshore Technology Conference, OTC, Houston, TX, OTC 6486 (1990). doi:10.4043/6486-MS

Schupp, J., et al.: Pipeline unburial behaviour in loose sand. In: Proceeding of the 25th International Conference of Offshore Mechanical Arctic Engineering, Hamburg, Paper OMAE2006-92542 (2006). doi:10.1115/OMAE2006-92542

Stone, K.J.L., Newson, T.A.: Uplift resistance of buried pipelines: an investigation of scale effects in model tests. In: Ng, C.W.W., Zhang, L.M., Wang, Y.H. (eds.) Proceeding of the 6th International Conference on Physical Modelling in Geotechnics, vol. 1, pp. 741–746. Taylor & Francis Group, London (2006)

Sun, J., Jukes, P.: Upheaval buckling analysis of partially buried pipeline subjected to high pressure and high temperature. In: Proceedings of the ASME 2011 30th International Conference on Ocean, Offshore and Arctic Engineering, OMAE2011-49498. ASME (2011). doi:10.1115/OMAE2011-49498

Berghe, V., et al.: Pipeline uplift mechanisms using finite element analysis. In: Proceedings of the 16th International Conference of Soil Mechanics and Foundation Engineering, pp. 1801–1804. Millpress Science Publishers, Fugro Engineers SA, Brussels (2005)

Vermeer, P.A., Sutjiadi, W.: The uplift resistance of shallow embedded anchors. In: Proceedings of the 11th Conference of Soil Mechanics and Foundation Engineering, San Francisco, vol. 3, pp. 1635–1638 (1985)

Wang, J., et al.: Mobilization distance for upheaval buckling of shallowly buried pipelines. J. Pipeline Syst. Eng. Pract. (2012). ASCE. doi:10.1061/(ASCE)PS.1949-1204.0000099

White, D.J., et al.: Centrifuge modelling of upheaval buckling in sand. Int. J. Phys. Model. Geotech. (2001). doi:10.1680/ijpmg.2001.010202

Yimsiri, S., et al.: Lateral and upward soil-pipeline interactions in sand for deep embedment conditions. J. Geotech. Geoenvironmental Eng. (2004). ASCE. doi:10.1061/ASCE1090-02412008134:2154

Zeng, X., et al.: Critical upheaval buckling forces of imperfect pipelines. J. Appl. Ocean Res. (2014). Elsevier. doi:10.1016/j.apor.2014.01.001

Experimental Investigation of Settlement Induced Bending Moments on Pile Supported T-Walls

Panagiota Kokkali[1(✉)], Tarek Abdoun[2], and Anthony Tessari[3]

[1] WSP|Parsons Brincherhoff, New York, NY, USA
kokkalip@pbworld.com
[2] Department of Civil and Environmental Engineering, Rensselaer Polytechnic Institute, Troy, NY, USA
[3] Department of Civil, Structural and Environmental Engineering, University at Buffalo, Buffalo, NY, USA

Abstract. T-Walls are pile-supported concrete floodwalls that are part of the New Orleans flood protection system. They are typically founded on long concrete or steel battered piles and include a sheet pile cut off for seepage and gradient control. Since the soils in Southern Louisiana contain soft and compressible silt and clay, these floodwalls need to be designed to resist not only flood loads but also loads produced by consolidation of soft foundation soil under the weight of new fill placed on or near a T-Wall during construction. This is commonly referred to as downdrag and induces additional bending moments on the piles. It is important that the settlement induced bending moments are assessed and understood to ensure that they do not exceed allowable limits. In an attempt to develop a comprehensive T-Wall design procedure that appropriately considers field conditions, the United States Army Corps of Engineers (USACE) New Orleans District partnered with Rensselaer Polytechnic Institute (RPI) and Virginia Polytechnic Institute and State University and jointly developed a research program. A series of centrifuge tests was conducted at the Centrifuge Facility at RPI in order to gain insight into the mechanisms and magnitudes of downdrag-induced bending moments and provide reliable data sets for validation of numerical models. Different pile configurations and fill loading conditions were tested in an attempt to isolate the key parameters of the system response to downdrag forces. This paper describes the experimental methodology and discusses data from two centrifuge tests in view of the effect of an additional row of piles on the Protected Side of the system.

1 Introduction

Following Hurricane Katrina, the United States Army Corps of Engineers performed an in depth evaluation of the hurricane protection system in New Orleans and Southeast Louisiana. I-Walls were soon deemed unsuitable for storm surge greater than 1.2 m. However, T-Walls proved to be an extremely robust alternative (IPET 2006; Brandon et al. 2008; Duncan et al. 2008; Sasanakul et al. 2008; Sills et al. 2008; Ubilla et al. 2008; USACE 2012). T-Walls are typically founded on long concrete or steel battered

T. Abdoun and S. Elfass (eds.), *Soil Dynamics and Soil-Structure Interaction for Resilient Infrastructure*, Sustainable Civil Infrastructures, DOI 10.1007/978-3-319-63543-9_8

piles and include a sheet pile cut off for seepage and gradient control. Figure 1 shows a typical section of a T-Wall. After Katrina, the USACE and RPI performed a comprehensive experimental investigation of T-Wall systems, focusing on the effect and magnitude of the battered piles on the stabilization of the foundation soil due to storm surge hydraulic loading and the impact of the sheet pile wall on the stability of the system (Tessari 2012). The findings from this experimental investigation were later incorporated into the T-wall design procedures (USACE 2012).

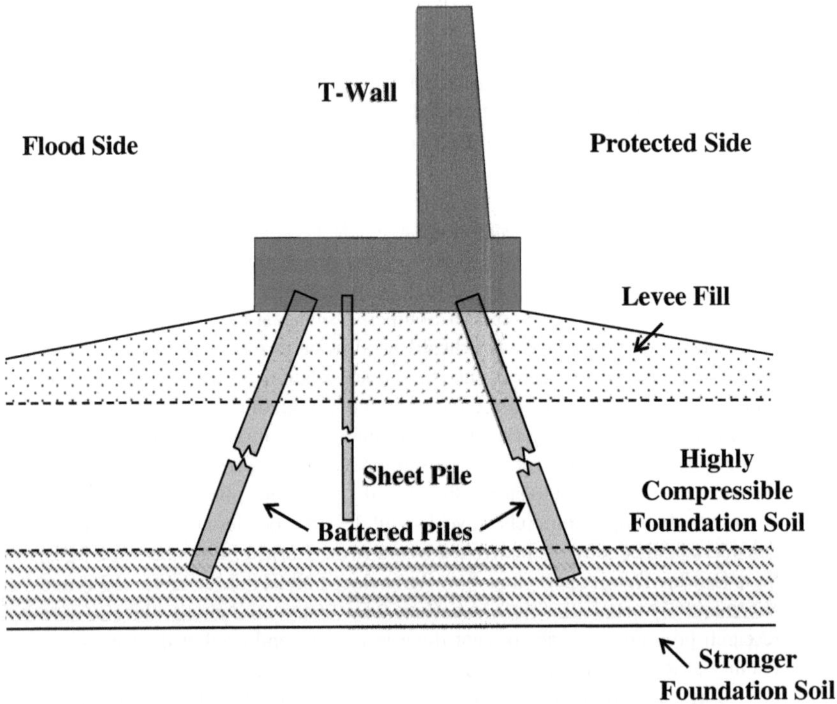

Fig. 1. Typical section of a T-Wall supported by battered piles

However, since the soils in Southern Louisiana contain soft and compressible silt and clay, these floodwalls need to be designed to resist not only the flood loads but also the loads produced by consolidation of the soft foundation soil under the weight of a levee embankment. The downdrag forces may lead to premature flexural failure and prohibit the system from retaining the designed level of storm floodwater. Numerical investigations have already been performed to represent the effects of settlement on pile-supported T-Walls (Isenhower et al. 2014; Reese and Associates 2011; McGuire et al. 2012). In an attempt to develop a comprehensive T-Wall design procedure that accurately represents the field conditions, the New Orleans District partnered with RPI and Virginia Polytechnic Institute and State University. A testing program was jointly developed to perform centrifuge tests that would provide insight into the mechanisms and magnitudes of downdrag-induced bending moments and reliable data sets for validation of numerical models. The proposed program consisted of eight centrifuge trials that were performed at the Center for Earthquake Engineering Simulation at Rensselaer. Different pile configurations and fill loading conditions were tested in an attempt to isolate the key parameters of the system response to downdrag forces. This paper describes the experimental methodology and discusses data from two centrifuge tests in view of the effect of an additional row of piles on the Protected Side of the system.

2 Experimental Methodology

The experimental series consisted of eight distinct trials focusing on different pile configurations (rows of piles, spacing, length) and fill loading conditions. The primary objective of this project was to isolate the effects of downdrag loading on the battered piles; therefore, the experimental trials did not include flood loading. Due to the inherent constraints of centrifuge modeling, several compromises had to been made in the design and execution of the centrifuge tests. Even though the tests were different enough from real conditions, they still provided the critical information for mechanisms observed in the field and can be used as the basis for validation of numerical models.

The centrifuge trials were modeled after the configuration of Fig. 1. Each model was comprised of a sand drainage layer at the bottom of the model container, a clay layer that represented the highly compressible foundation soil, and a sand embankment. The experimental configurations discussed in this paper are shown in Fig. 2. The fill load consisted of symmetrical embankments of limited extent, constructed out of Nevada sand. The T-Wall had a fixed geometry and was designed so that it could be supported by two or three rows of piles. In the first configuration, Run 2, the T-Wall was supported by two rows of piles. In the second configuration, Run 4, the T-Wall was supported by an additional row of piles on the Protected Side.

The experimental methodology for this testing series with respect to the selection of the structural elements of the system and the scaling constraints are thoroughly discussed in Kokkali et al. (2016). The testing g-level was iteratively selected to be 70 g after taking into consideration the critical pile properties (width, flexural rigidity, axial stiffness, and pile length) and realistic consolidation times. Since aluminum has a unit weight similar to concrete, the T-Wall foundation was machined out of aluminum. A commercially available brass rectangular tube was used to model the battered piles with comparable properties to an HP 14 $\times$ 73, as shown in Table 1.

All piles extended through the 21 m (70 ft) of kaolinite clay at a 3 to 1 slope (3V: 1H). The piles were pinned at the bottom ends, essentially simulating infinitely long elements. Since the maximum effect of downdrag has been estimated to occur at depths shallower than 21 m, the selected pile lengths and bottom pin connections were considered appropriate for centrifuge simulation purposes. Piles in the field are embedded in the T-Wall producing an intermediate fixation between fixed and pinned conditions. Given the physical dimensions necessary for a three-leg arrangement, a pinned connection at the T-Wall foundation level would have been difficult to produce; therefore, a fixed connection was preferred. In order to simulate the correct soil consolidation near the soil-pile boundary, protect the instrumented piles and ensure the pile alignment, the

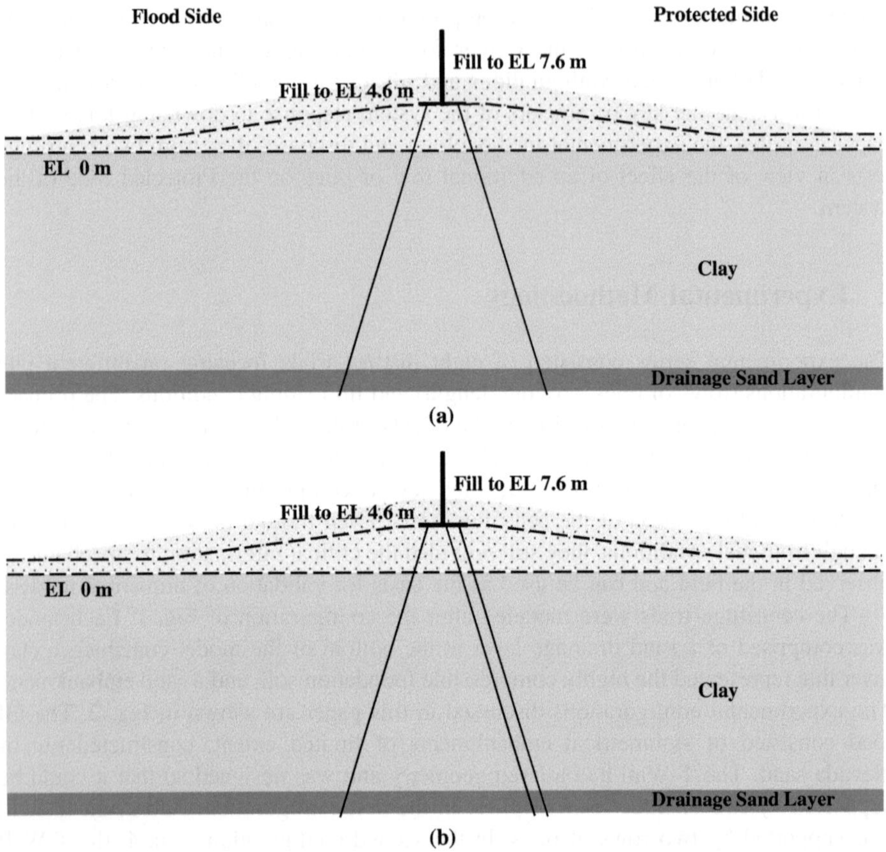

Fig. 2. The experimental configurations discussed in this paper: (a) Run 2, (b) Run 4

Table 1. Properties of field and prototype battered piles

Properties	HP 14 × 73	Prototype
Material	Steel	Brass
Section (m × m)	0.37 × 0.35	0.22 × 0.45
Bending stiffness EI (Nm^2)	6.07 E+07	7.68 E+07
Axial stiffness EA/L (N/m)	2.76 E+09	3.18 E+09

piles were installed and pin connected to the bottom of the container prior to the construction of each model (at 1 g). They were free to rotate to their final locations during the initial consolidation phase and a pile guide was used to assure proper alignment in two planes.

In order to capture the response of the system and characterize the soil deposit several sensors were installed in the model: (a) strain gauges on the piles, (b) pore pressure transducers in the clay layer, (c) displacement tracking targets on the transparent boundary of the container, (d) bender elements in the clay layer to obtain the shear wave velocity of the clay, and (e) linear variable differential transformers (LVDTs) to record the settlement at the soil surface. Figure 3 provides the typical instrumentation layout for a T-Wall test. The exact locations of the sensors varied upon the tested configuration.

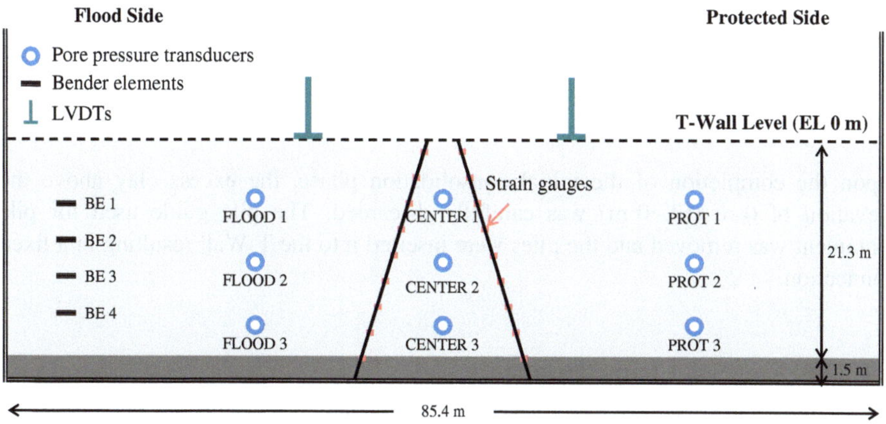

Fig. 3. The experimental configurations discussed in this paper: (a) Run 2, (b) Run 4

3 Testing Sequence

The testing sequence for each run included the model construction, the initial clay consolidation, in-flight soil characterization, fill loading testing, and post experiment soil characterization and model deconstruction.

3.1 Model Construction

Each model consisted of a sand layer for drainage purposes at the bottom of the container, a clay layer, and a sand embankment. Prior to the model construction, the battered piles were installed and aligned in the model container using a pile guide. Kaolinite clay was mixed at 70% initial water content (LL = 60) and installed in the container on top of the sand drainage layer. It extended above the desired elevation (EL 0 m) to account for settlement that would occur during the initial consolidation phase. A sand surcharge was placed on top of the clay layer to provide additional surcharge and prevent clay desiccation.

3.2 Initial Consolidation of the Clay Layer

The initial consolidation phase required several days of testing to simulate a representative amount of prototype years in the field and was performed in steps. The objective of the in-flight consolidation was to produce a pre-consolidation pressure profile that was close to normally consolidated conditions when the consolidation surcharge and the excess clay were removed. This facilitated both physical and numerical modeling rather than attempting to achieve a specific shear strength profile. During this step, the piles were free to rotate at the top, while they slowly moved to their final locations. The pore water pressure in the clay layer and the surface settlement were continuously monitored. The readings from these sensors indicated when the initial consolidation phase should be stopped.

3.3 T-Wall Installation

Upon the completion of the initial consolidation phase, the excess clay above the elevation of 0 m (EL 0 m) was carefully discarded. The pile guide used for pile alignment was removed and the piles were inserted into the T-Wall resulting in a fixed connection.

3.4 In-Flight Soil Characterization

In most tests, in-flight T-Bar testing was performed in order to determine the initial in-situ undrained shear strength of the clay layer. T-Bar testing was not performed during Run 2; therefore, S_u profiles are not presented in this paper. The shear strength testing protocol and method of analysis are described in Kokkali et al. (2016).

3.5 Establish no Load Conditions - Nulling

The model was spun to the design g level with no embankment in place to establish the reference no load conditions. This phase is referred to as 'nulling' in the ensuing section.

3.6 Fill Loading

Field installations utilize surcharge embankment material that is similar to the foundation soil. This would be nearly impossible to replicate in the centrifuge and it would create several uncertainties. In order to facilitate both the construction of the models and the ability to model the geometry numerically, sand was selected for the surcharge. The sand could be easily placed in a consistent and repeatable manner at a unit weight similar to that found in a field installation. The levee embankment was built and tested in two phases during Runs 2 and 4: (a) fill to the T-Wall base at EL 4.6 m, and (b) fill to EL 7.6 m.

3.7 Post Experiment Soil Characterization and Model Deconstruction

In most Runs, the post experiment soil characterization included additional in-flight T-Bar testing. During the model excavation, water content samples were selected and the final locations of the sensors were recorded. Any observations about the deformed model were logged to supplement the data reduction.

4 Centrifuge Test Results

Table 2 provides a summary of the testing stages, including consolidation times for each phase of interest. Since Run 4 was performed later than Run 2 some of the steps were slightly modified to address certain issues that arose during the first trials. In that context, Run 4 initial consolidation was longer in order to achieve a higher degree of consolidation for a thicker initial clay layer. However, since the majority of the clay settlement occurred during the first hours of consolidation, the soil profile generated during Run 4 was not significantly different from the profile generated during Run 2. The pore water pressure and the shear wave velocity measurements at the beginning of the fill loading sequence shown in Fig. 4 indicate that the clay profiles were comparable in terms of degree of consolidation and stiffness.

Table 2. Summary of testing phases for Run 2 and 4

Phase	Run 2	Run 4
Initial consolidation	8.9 years	13.4 years
Nulling	Establish reference conditions	
Fill to EL 4.6 m	6.7 years	5.6 years
Fill to EL 7.6 m	6.7 years	5.6 years

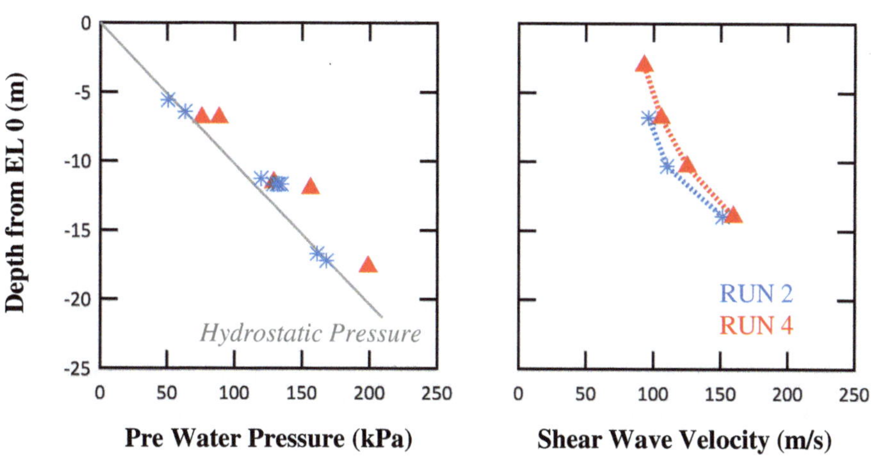

Fig. 4. Typical instrumentation layout for a T-Wall experiment

This section focuses on the bending moments induced on the battered piles as the foundation clay consolidated under the weight of the embankment. Data from instrumented piles located in the middle of the test container are presented in Figs. 6 and 7. The strain gauge readings on the piles were used to calculate the bending moments along the piles by interpolating between neighboring locations. Figure 5 shows the bending moment sign convention. The pile on the left is referred to as 'Flood Side pile FS' and the pile to the right as 'Protected Side pile PS'.

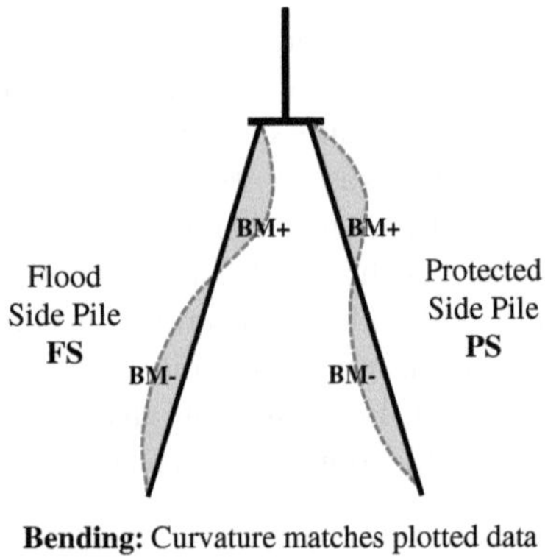

Fig. 5. Bending moment sign convention

Figure 6 shows the bending moments along the flood and the protected side piles developed by the end of each testing phase (Nulling, Fill to EL 4.6 m, and Fill to EL 7.6 m). Depth zero on the vertical axis corresponds to EL 0 m, while the T-Wall rests at 4.6 m above EL 0 m. As anticipated, the consolidation under the new fill led to considerable increase in the bending moments in both experiments. The maximum moments were registered at the fixed connections and between EL 0 to −5 m. The magnitude and location of maximum settlement induced bending moments are critical since they define the remaining capacity of the system to resist flood loading.

Overall, higher bending moments were observed during Run 2. As shown in Fig. 6, the presence of the additional row of piles on the Protected Side alleviated the Run 4 piles. The bending moment distributions recorded during Run 2 were not as symmetrical as expected due to an experimental error. Due to discrepancies in the final position of the piles, the Protected Side piles of Run 2 were more prone to downdrag effects. In order to better compare the two runs, the initial loading on the piles (nulling) was subtracted from the bending moments at the end of each loading phase and is shown in Fig. 7. The effect of the third pile on the Protected Side of Run 4 is more evident in these graphs. While the conditions on the Flood Side piles were still quite similar during both Runs, the Protected Side piles of Run 4 carried about half of the settlement induced bending moments measured in Run 2. This response was directly linked to the observed settlement during the tests. As qualitatively indicated in Fig. 8, the extent of deformation around the Run 2 piles was higher. In other words, the

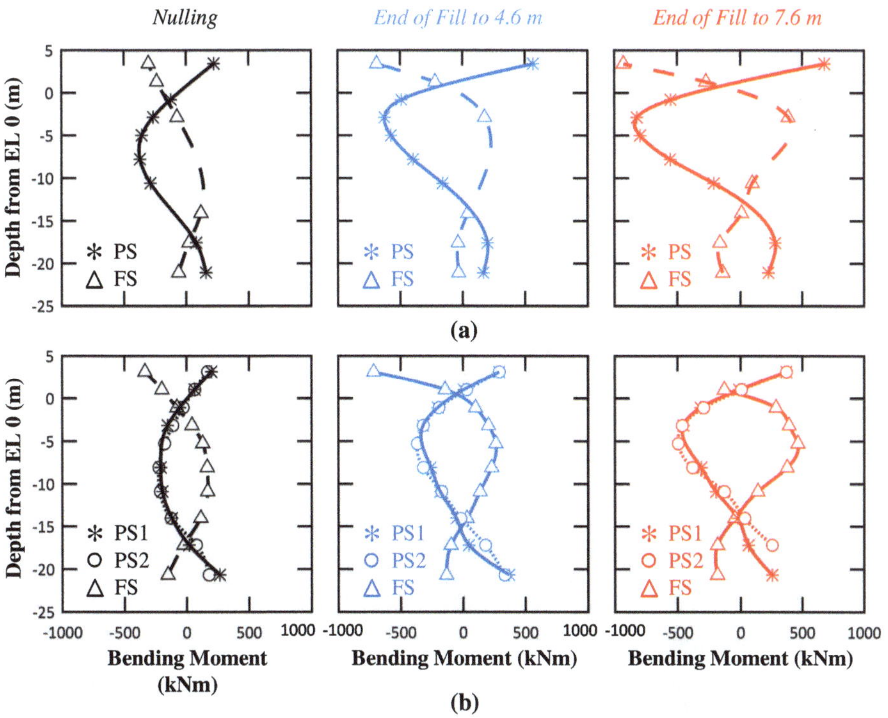

Fig. 6. Bending moments developed along the battered piles at the end of each loading phase (a) Run 2, (b) Run 4

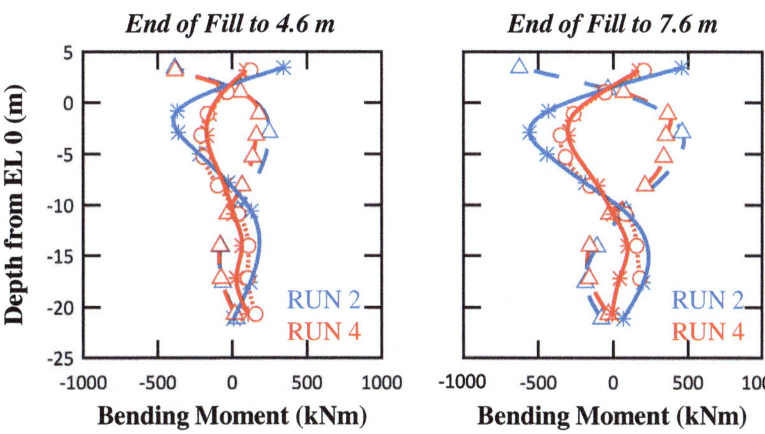

Fig. 7. Comparison of settlement induced bending moments developed along the battered piles at the end of the loading phases during the two runs

Run 2

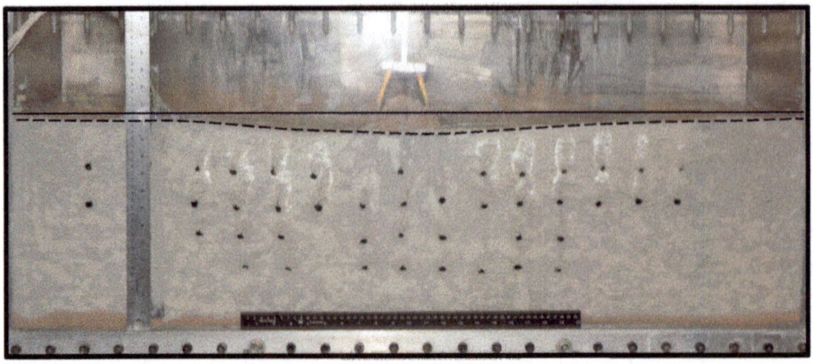

(a)

Run 4

(b)

Fig. 8. Deformed models after removing the sand fill: (a) Run 2, (b) Run 4

presence of the third row of piles in Run 4 not only led to redistribution of the loads among three rows of piles instead of two, but also reduced the settlement around the piles, therefore, reduced the additional loads on the piles.

5 Conclusions

A centrifuge test series was performed at the Center for Earthquake Engineering Simulation at RPI in order to study the settlement induced bending moments on battered piles that support floodwalls commonly built in New Orleans. Two centrifuge tests were presented in this paper simulating the consolidation of a soft foundation soil under a symmetrical embankment of limited extent. The centrifuge trials included a symmetrical two-leg configuration and an asymmetrical three-leg configuration. At the first stage of the fill loading phase the embankment was built up to the T-Wall level and during the second stage additional fill was added.

The evolution of bending moments as the foundation soil consolidated under the embankment weight was recorded and indicated significant increase in both centrifuge trials with maximum moments observed at the fixed connection with the T-Wall and around the top third of the piles length. The bending moments developed on the battered piles of the three-leg configuration were overall lower. Specifically, the presence of the second row of piles on the protected side led to decrease in the settlement induced bending on the protected side, while the conditions remained the same on the flood side.

The problem has been further explored experimentally including various pile configurations and fill loading cases. Additional data and in depth conclusions will be presented in future publications.

Acknowledgements. The financial support for this paper has been provided by the United States Army Corps of Engineers. The authors would also like to thank the staff and undergraduate researchers of the Center for Earthquake Engineering Simulation at RPI for their contribution to this project.

References

Brandon, T.L., Wright, S.G., Duncan, J.M.: Analysis of the stability of I-Walls with gaps between the I-Wall and the levee fill. J. Geotech. Geoenviron. Eng. **134**(5), 692–700 (2008)

Duncan, J.M., Brandon, T.L., Wright, S.G., Vroman, N.: Stability of I-walls in new orleans during hurricane katrina. J. Geotech. Geoenviron. Eng. **134**(5), 681–691 (2008)

Interagency Performance Evaluation Taskforce (IPET): Performance evaluation of the New Orleans and Southeast Louisiana Hurricane Protection system. Draft final Report, U.S. Army Corps of Engineers (2006)

Isenhower, W., Vasquez, L., Wang, S.: Analysis of settlement-induced bending moments in battered piles. In: From Soil Behavior Fundamentals to Innovations in Geotechnical Engineering, pp. 497–511 (2014)

Kokkali, P., Tessari, A., Abdoun, T., Varuso, R., Johnson, J., Filz, G., Reeb, A.: Settlement induced bending moments on pile founded floodwalls. Under Revision in the Int. J. Phys. Model. Geotech. (2016)

McGuire, M., Filz, G., Navin, M., Hokens, K.: LPILE Method for Evaluating Bending Moments in Batter Piles due to Ground Settlement for Pile-Supported Floodwalls in New Orleans and Vicinity, Rep. prepared for USACE Contract No. W912P8-07-D-0062 (2012)

Reese, L.C., Associates: Evaluation of Bending in Batter Piles due to Settlement. Report prepared for FFEB Joint Venture, LLC, Lymon C. Reese & Associates, Austin, TX (2011)

Sasanakul, I., Vanadit-Ellis, W., Sharp, M., Abdoun, T., Ubilla, J., Steedman, S., Stone, K.: New Orleans levee system performance during Hurricane Katrina: 17th Street Canal and Orleans Canal North. J. Geotech. Geoenviron. Eng. **134**(5), 657–667 (2008)

Sills, G.L., Vroman, N.D., Wahl, R.E., Schwanz, N.T.: Overview of New Orleans levee failures: lessons learned and their impact on national levee design and assessment. J. Geotech. Geoenviron. Eng. **134**(5), 556–565 (2008)

Tessari, A.: Centrifuge modeling of the effects of natural hazards on pile-founded concrete floodwalls. Ph.D. Dissertation, Rensselaer Polytechnic Institute, Troy, NY (2012)

Ubilla, J., Abdoun, T., Sasanakul, I., Sharp, M., Steedman, S., Vanadit-Ellis, W., Zimmie, T.: New Orleans levee system performance during hurricane Katrina: London Avenue and Orleans canal south. J. Geotech. Geoenviron. Eng. **134**(5), 668–680 (2008)

USACE: Hurricane and Storm Damage Risk Reduction System Design Guidelines, Interim with revisions through June 2012, USACE, New Orleans (2012)

Soil-Pile-Structure Interaction Evidences from Scaled 1-g model

M.G. Durante[1(✉)], L. Di Sarno[2], George Mylonakis[3], Colin A. Taylor[3], and A.L. Simonelli[2]

[1] Department of Civil and Environmental Engineering, University of California, Los Angeles, CA, USA
mgdurante@ucla.edu
[2] Dept. of Engineering, Università degli Studi del Sannio, Benevento, Italy
[3] Department of Civil Engineering, University of Bristol, Bristol, UK

Abstract. The seismic soil-pile-structure interaction (SPSI) is a complex mechanism that is usually considered formed by the combination of kinematic and inertial interaction. Both mechanisms generate additional forces on the system related to the stiffness contrast between the soil and the foundation for the kinematic interaction and to the soil response around the foundation due to the inertial contribution of the superstructure for the so-called inertial interaction. While the mechanism of each of these contributions is clear, their combination is still under investigation due to number of parameters involved (i.e. dynamic characteristics of both system and input). An effective way to study this combination is the analysis of actual data on real structures. Due to the fact that these data are hard to find, usually the response of physical scaled models on 1-g and n-g devices are investigated. In this connection, this paper presents some results from an extensive 1-g shaking table testing activity. The scaled physical model is formed by a group of five piles embedded in a by-layer deposit of dry sands with an oscillator connected to the piles through different kind of foundation systems. More specifically, the attention is focused on both pile and structural response when the oscillator is connected to a small group of three piles by means of a stiff foundation. The analysis of the experimental data enhances the role of the resonance between the soil-structure system and the input waves in the general behavior of both structure and piles.

1 Introduction

The seismic Soil-Pile-Structure interaction (SPSI) is the mutual interaction among soil, foundation and superstructure during seismic loads. The complex phenomenon is usually considered as the combination of kinematic and inertial interaction (Fig. 1). The kinematic interaction is related to the influence of the soil in the pile response under the upcoming seismic waves, while the inertial interaction is related to the influence of the superstructure to the pile response. The combination of these two mechanisms is a crucial point for the evaluation of the general response of the system and it is not trivial. The main effects of the SPSI on a structure are related to changing of the dynamic properties of the system in terms of natural frequencies and damping ratio.

T. Abdoun and S. Elfass (eds.), *Soil Dynamics and Soil-Structure Interaction for Resilient Infrastructure*, Sustainable Civil Infrastructures, DOI 10.1007/978-3-319-63543-9_9

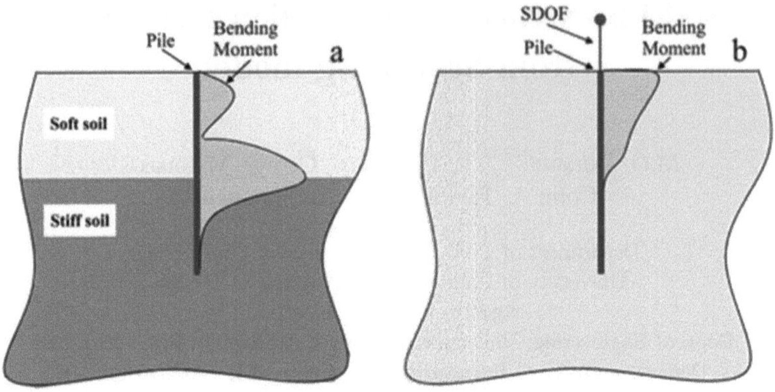

Fig. 1. (a) Kinematic and (b) inertial bending moments (qualitative patterns) (modified from Durante 2015)

In the last decades a lot of researchers focused their attention on this phenomenon analyzing the main parameters that affect its behavior. An extensive historical review is reported in Kausel 2010. The role of the SPSI on the seismic performance of the structure is usually considered beneficial (i.e. Jennings and Bielak 1973; Veletsos and Meek 1974; Bielak 1975; Roesset 1980; Luco 1982; Wolf 1985; Gazetas and Mylonakis 1998; Stewart and Fenves 1998, Stewart et al. 1999a and 1999b), but it can become detrimental under specific geotechnical and seismic conditions (Mylonakis and Gazetas 2000).

An effective way to study this phenomenon could be based on the analysis of real data collected from real structures or scaled ones. The latter can be investigated by means of experimental campaigns in n-g or 1-g devices (i.e. Meymand 1998; Muir Wood et al. 2002; Pitilakis et al. 2008; Moccia 2009; Chidichimo 2014; Durante 2015; Durante et al. 2016 among others).

In this paper the SPSI is investigated by means of a series of high quality of 1-g shaking table tests carried out at the Bristol Laboratory for Advanced Dynamics Engineering (BLADE) of the University of Bristol (UK), within the Framework of the Seismic Engineering Research Infrastructures for European Synergies (SERIES), which was funded by the 7th Programme of the European Commission. The scaled physical model is formed by a group of five pile embedded in a by-layer deposit of dry sands with an oscillator connected to the piles through different kind of foundation systems. The results here reported refer to the configuration with the oscillator resting on a foundation connecting three piles. The analysis of these data shows that an important role in the system response is played by the resonance conditions that develop during the motion.

2 Shaking Table Tests

The experimental campaign carried out at the BLADE was performed in two different phases using the six degrees of freedom shaking table. The 1-g shaking table is a 3 m x 3 m cast-aluminum platform powered by eight hydraulic actuators. In order to reduce the influence of the soil container on the soil response, the container used is an equivalent shear beam container formed by eight rectangular aluminum rings alternated by rubber section. The shear-stack is characterized by resonant frequency sufficiently different from the values obtained when it is filled with soil (Crewe et al. 1995).

The physical model analyzed during the experimental campaign is schematically reported in Fig. 2. The main properties of the model were obtained applying scaling laws to a prototype: here they are based on geometrical similarity, assuming the fundamental scale factor for length equal to 37.5 (Muir Wood et al. 2002).

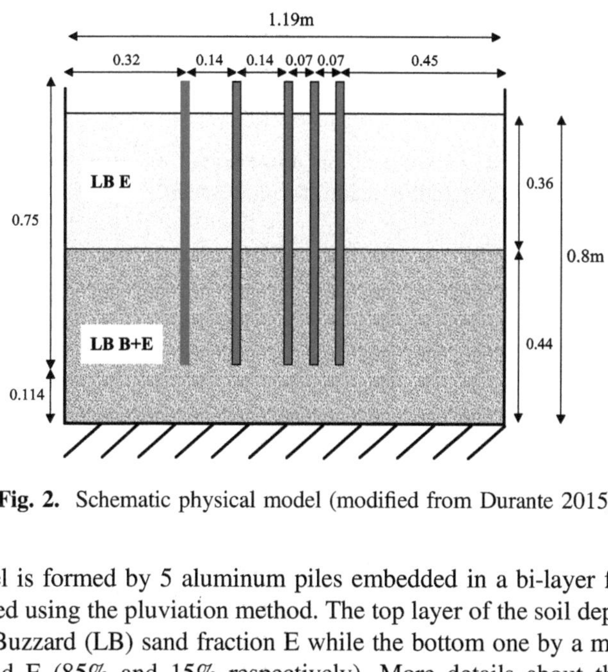

Fig. 2. Schematic physical model (modified from Durante 2015)

The model is formed by 5 aluminum piles embedded in a bi-layer formed by dry sands deposited using the pluviation method. The top layer of the soil deposit is formed by Leighton Buzzard (LB) sand fraction E while the bottom one by a mix of LB sand fraction B and E (85% and 15% respectively). More details about the mechanical properties of the soil deposit are reported in Table 1. Each pile is modelled using a 750 mm long alloy aluminum tube (thickness 0.71 mm) with the external diameter equal to 22.23 mm. Five different superstructures are considered in this work, made of five different masses attached to the top of the same column (cross Sect. 3 mm × 12 mm, height 100 mm) The dynamic properties of the fixed base oscillators were experimentally determined attaching them directly to the shaking table (Table 2).

Table 1. Soil layer properties (modified from Durante 2015)

Soil layers	Thickness H [mm]	Dry unit weight γ_d [kN/m^3]		Shear wave velocity V_s [m/s]		V_{s2}/V_{s1}	
		Phase I	Phase II	Phase I	Phase II	Phase I	Phase II
Top LB (E)	340	13.63	13.13	51	54	1.59	1.57
Bottom LB (E + B)	460	17.46	17.92	81	85		

Table 2. Experimental frequency and damping of fixed-base oscillators

Mass [gr]	Frequency [Hz]	Damping ratio [%]
125	30.5	1.2
175	26.5	0.9
275	20.5	1.4
475	15.0	1.2
975	10.4	1.5

The dynamic input motions used during the experimental campaign can be summarized as follows:

- *White noise excitation:* it is a random noise signal of bandwidth 0–100 Hz with fixed value of peak acceleration, useful for the dynamic identification of the system and its parts;
- *Harmonic excitation:* it is a sinedwell input motion. The frequency range of the input is 5–50 Hz while the acceleration range is 0.01 g – 0.13 g;
- *Earthquake excitation:* the records used are selected from the SISMA database (Scasserra et al. 2008) and modified according to the adopted scaling laws (Muir Wood et al. 2002). The three records selected are Tolmezzo from the Friuli 1976 earthquake, Sturno from the Irpinia 1980 earthquake and Nocera Umbra Biscontini from the Umbria-Marche 1997 event.

The sample model was monitored with 62 devices consisting of 18 accelerometers (on shaking table, shear-stack, soil column, pile heads and oscillator), 36 strain-gauges for the evaluation of bending moment and axial forces of the piles and 8 linear variable displacement transducers in the vertical and horizontal directions (on shear-stack and piles head).

3 Test Results

Typical results of the shaking table tests described in the previous section are presented herein. The effects of the soil-pile-structure interaction are presented referring to the Short Cap configuration with the oscillator (SC + SDOF, Fig. 3) that consists of an oscillator connected to the piles by means of a rigid foundation that connects three piles. Among the available tests, the harmonic input motion ones are selected for this work. More specifically results refer to three sets of data (Table 3) characterized by the same input motion (namely 10, 15 and 20 Hz sinedwell excitations) but 5 different oscillator masses (125 g, 175 g, 275 g, 475 g and 975 g).

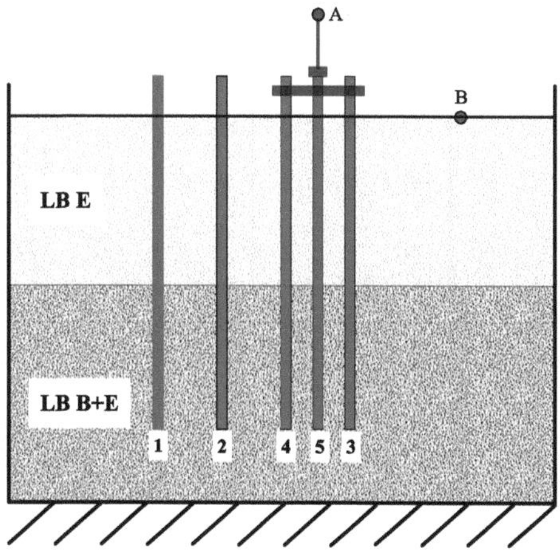

Fig. 3. Short-cap connection with oscillator configuration (SC + SDOF, modified from Durante 2015)

Table 3. Harmonic input motion considered

	Frequency [Hz]	Input acceleration [g]
SET 1	10	0.15
SET 2	15	0.10
SET 3	20	0.10

Test results are presented focusing on two different effects of SPSI on the system response: first on structural response, then on pile response.

4 Structural Response

The structural response is evaluated computing the variation of the natural frequency and damping ratio of the system with respect to the fixed base condition. Towards this aim, for each test, the transfer function (TF), i.e. the ratio of the FFT of the accelerogram at the top of the oscillator and the counterpart at the soil surface (A and B in Fig. 3, respectively) is computed.

The computed transfer functions for all the masses are shown in Fig. 4: the natural frequency of the oscillator decreases with mass. A preliminary visual inspection of Fig. 4 suggests a trend of decreasing amplitude ratio as the natural frequency of the system approaches the input frequency, namely the oscillator with the mass of 975 g for SET 1 (f_{SSI} = 10 Hz), mass of 475 g for SET 2 (f_{SSI} = 14.2 Hz) and the mass of 275 g for SET 3 (f_{SSI} = 19.6 Hz). This trend can be related to a strong damping effect, that possibly dominates the behavior of the system under resonance conditions. This outcome should be further verified using additional data and/or numerical simulations. The variation of the dynamic parameters of the flexible base systems are reported in Table 4 and Fig. 5. As far as the period elongation is concerned, it is evident that, even

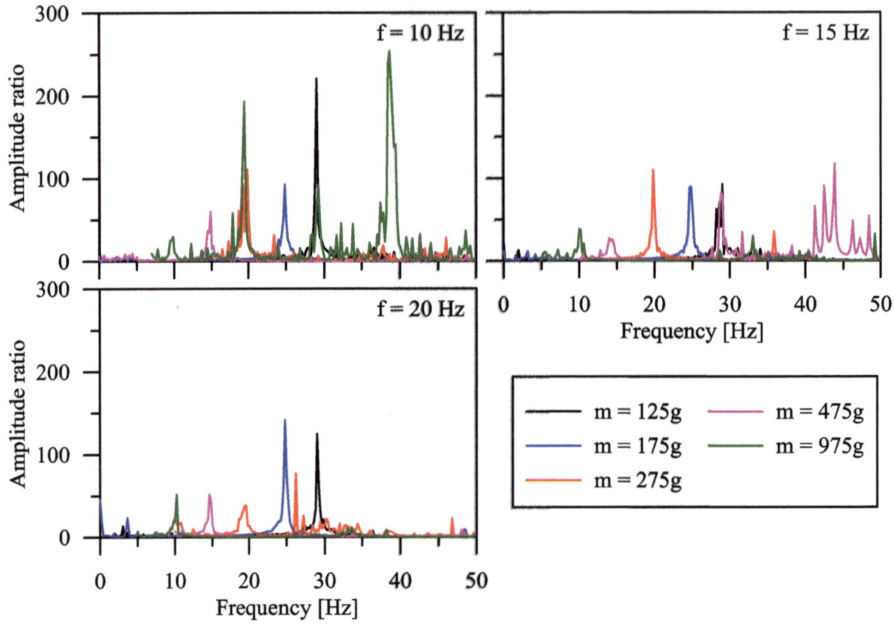

Fig. 4. Transfer functions for the three set of data considered

Table 4. Dynamic properties of selected tests

	m[g]	1/σ	T_{SSI}/T_{fix}	D_{SSI}/D_{fix}
SET 1 (f = 10 Hz, a = 0.15 g)	125	0.095	1.051	0.242
	175	0.080	1.069	0.744
	275	0.061	1.035	0.407
	475	0.043	1.014	0.846
	975	0.032	1.061	1.400
SET 2 (f = 15 Hz, a = 0.10 g)	125	0.094	1.051	0.583
	175	0.080	1.069	0.778
	275	0.063	1.035	0.407
	475	0.043	1.056	1.750
	975	0.031	1.040	1.067
SET 3 (f = 20 Hz, a = 0.15 g)	125	0.093	1.051	0.417
	175	0.081	1.069	0.489
	275	0.062	1.046	1.143
	475	0.046	1.027	1.042
	975	0.032	1.020	1.667

if there are resonant effects on the system in the tests considered, its variation is negligible. Such structural response has been attributed to the particular restrain condition for the oscillator that is stiff enough to be assimilated to the fixed base condition. The dynamic properties of the systems are also plotted in Fig. 5 versus the so-called "wave parameter" ($1/\sigma = f_{fix}\ h/V_s$) that can be considered an index of the relative stiffness between the structure (of height h) and the soil. The shear wave velocity (V_s) is computed experimentally (Durante et al. 2015) and it is assumed equal to the mean value of the top layer of the deposit in which lies the active length of the pile.

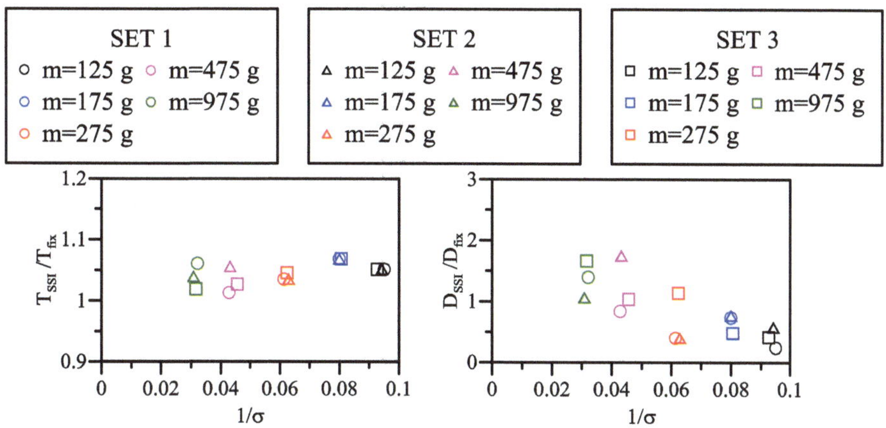

Fig. 5. Dynamic response of the system: period elongation (left) and damping ratio (right) versus wave parameters for the tests considered

5 Geotechnical Response

In this section the geotechnical response of the system is evaluated for each test, comparing the free-field and pile responses.

In each set of data, the free-field response (Fig. 6a) is evaluated comparing the absolute maximum acceleration versus depth along a vertical array located far enough from piles to be considered independent from the system configuration. As expected, the response of the soil is affected only by the characteristics of the input motion and not by the mass of the oscillator.

The bending moments at each monitored elevation are computed using the strain-gauges records; their envelope in terms of absolute values are reported in Figs. 6b and c versus depth for pile 4 and pile 5. For each set of data, the maximum bending moments at pile 4 and 5 heads are almost the same, due to the capability of the rigid connection at the pile top to redistribute the inertial and kinematic effects. The combination of these two mechanisms at the pile head is not a trivial task, but the experimental data show that the main contribution is due to the kinematic interaction, induced by the high stiffness of the connection. As observed in the previous section, also from a geotechnical point of view, there is one oscillator for each set that behaves differently.

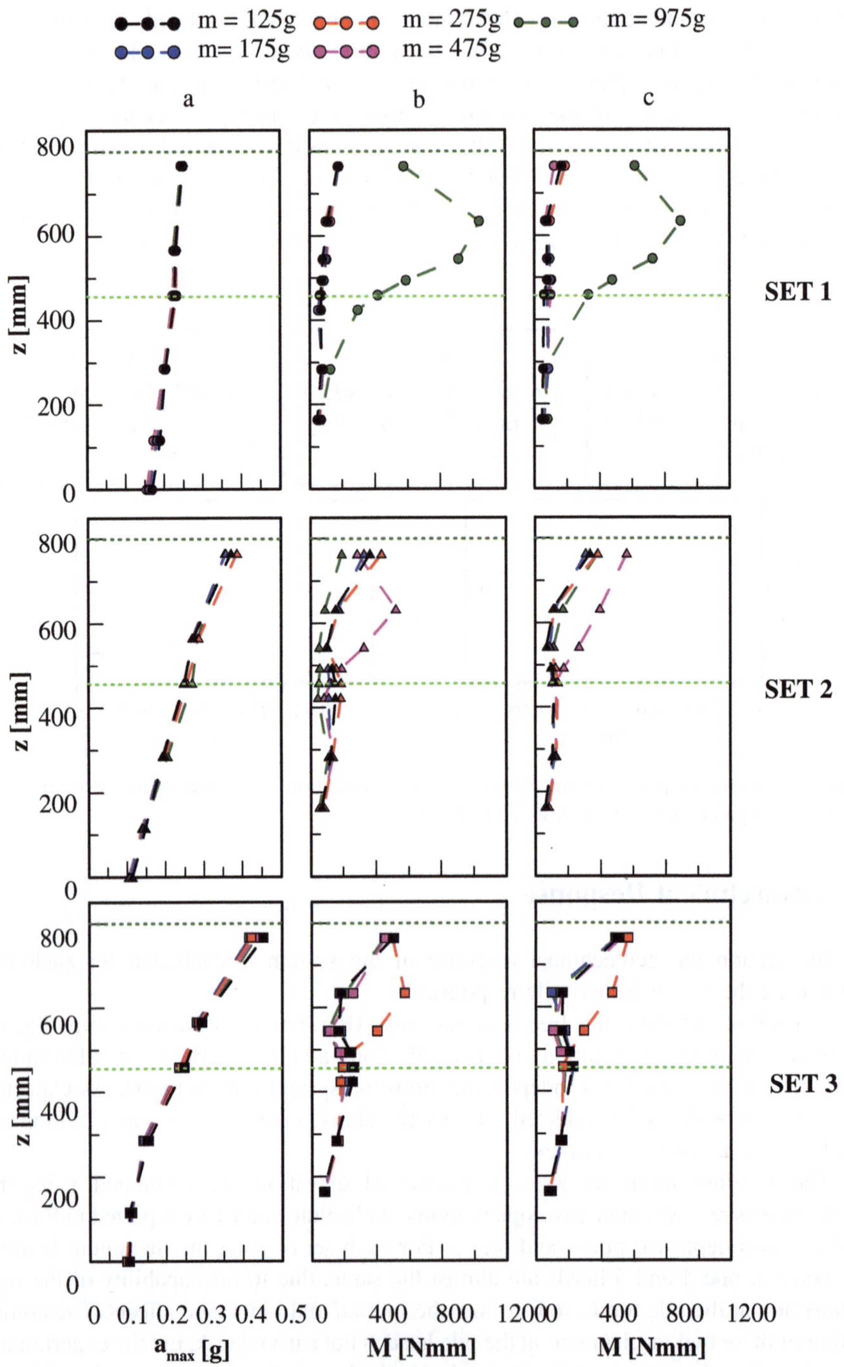

Fig. 6. Envelope of free-field accelerations (a) and bending moments along piles 4 (b) and 5 (c) for the tests considered

The envelopes of the absolute bending moment along both pile 4 and pile 5 show a different shape for the tests where the masses are in resonance with the input motion (mass of 975 g for SET 1, mass of 475 g for SET 2 and the mass of 275 g for SET 3).

6 Conclusions

The paper presented selected results from an extensive experimental campaign carried out at the Bristol Laboratory for Advanced Dynamic Engineering (BLADE) of the University of Bristol, within the Framework of the Seismic Engineering Research Infrastructures for European Synergies (SERIES).

The scaled physical model studied was tested on a 3 m x 3 m 1-g shaking table device. The model was formed by a group of five piles embedded in a deposit of dry sand with an oscillator connected to the piles through a foundation that involves a small group of three piles.

The soil-pile-structure interaction (SPSI) effects were analyzed considering three set of data, each one characterized by the same input motion (sinusoidal input) but 5 different oscillator masses, focusing the attention on both structural and geotechnical responses.

The analyses of the experimental data show that the resonance between the flexible base system and the input change the response of the structure, increasing the damping ratio and the response of the piles, changing the distribution of the bending moment along it. These outcomes are evident even if the period elongation of the system can be always considered negligible for this configuration, due to the significant stiffness of the connection among the piles: this connection practically reproduces the fixed base condition for the oscillator, inducing a significant kinematic interaction at the pile head.

Acknowledgments. The research leading to these results has received funding from the European Union Seventh Framework Programme (FP7/2007–2013) under grant agreement no. 227887, SERIES.

The financial support provided by the ReLUIS (TaskMT2) project funded by the Italian Civil Protection (Agreement No. AQDPC/ReLUIS 2014–2018) is also appreciated.

The authors would like to acknowledge all the contributors to the SERIES TA4 project, namely: Dr. Matthew Dietz and Dr. Luiza Dihoru from University of Bristol, Prof Subhamoy Bhattacharya from University of Surrey (formerly at University of Bristol), Dr Stefania Sica from University of Sannio, Prof. Gianni Dente, Dr. Roberto Cairo and Dr. Andrea Chidichimo from University of Calabria, Prof. Arezou Modaressi and Dr. Luìs A. Todo Bom from Ecole Centrale Paris, Prof. Amir M. Kaynia from Norwegian University of Science and Technology and Dr. George Anoyatis from University of the West of England (formerly at University of Patras).

References

Kausel, E.: Early history of soil–structure interaction. Soil Dyn. Earthq. Eng. **30**(9), 822–832 (2010). doi:10.1016/j.soildyn.2009.11.001. Special Issue in honour of Prof. Anestis Veletsos

Jennings, P.C., Bielak, J.: Dynamics of building-soil interaction. Bull. Seismol. Soc. Am. **63**(1), 9–48 (1973)

Veletsos, A.S., Meek, J.W.: Dynamic behavior of building-foundation systems. Earthq. Eng. Struct. Dyn. **3**(2), 121–138 (1974)

Bielak, J.: Dynamic behaviour of structures with embedded foundations. Earthq. Eng. Struct. Dyn. **3**(3), 259–274 (1975)

Roesset, J.M.: A review of soil-structure interaction. In: Johnson, J.J. (ed.) Soil-Structure Interaction: The Status of Current Analysis Methods and Research, US Nuclear Regulatory Commission and Lawrence Livermore Laboratory, Report No. NUREG/CR-1780 and UCRL-53011 (1980)

Luco, E.: Linear soil-structure interaction: a review. In: Earthquake Ground Motions and Effects on Structures, ASME. AMD, vol. 53, pp. 41–57 (1982)

Wolf, J.P.: Dynamic Soil-Structure Interaction. Prentice hall Inc., New York (1985)

Gazetas, G., Mylonakis, G.: Seismic soil-structure interaction: new evidence and emerging issues. In: Dakoulas, P., Yegian, M.K., Holtz, R.D. (eds.) Geotechnical Earthquake Engineering and Soil Dynamics III, ASCE, vol. II, pp. 1119–1174 (1998)

Stewart, J.P., Fenves, G.L.: System identification for evaluating soil-structure interaction effects in buildings from strong motion recordings. Earthq. Eng. Struct. Dyn. **27**(8), 869–885 (1998)

Stewart, J.P., Fenves, G.L., Seed, R.B.: Seismic soil-structure interaction in buildings. I: analytical methods. J. Geotech. Geoenvironmental Eng. **125**(1), 26–37 (1999a)

Stewart, J.P., Seed, R.B., Fenves, G.L.: Seismic soil-structure interaction in buildings. II: empirical findings. J. Geotech. Geoenvironmental Eng. **125**(1), 38–48 (1999b)

Mylonakis, G., Gazetas, G.: Seismic soil-structure interaction: beneficial or detrimental? J. Earthq. Eng. **4**(3), 277–301 (2000)

Meymand, P.J.: Shaking table scale model tests of nonlinear soil-pile-superstructure interaction in soft clay. Ph.D thesis, University of California, Berkeley (1998)

Muir Wood, D., Crewe, A., Taylor, C.A.: Shaking table testing of geotechnical models. UPMG-Int. J. Phys. Model. Geotech. **2**, 01–13 (2002)

Pitilakis, D., Dietz, M., Wood, D.M., Clouteau, D., Modaressi, A.: Numerical simulation of dynamic soil–structure interaction in shaking table testing. Soil Dyn. Earthq. Eng. **28**, 453–467 (2008)

Moccia, F.: Seismic soil pile interaction: experimental evidence. Ph.D thesis, Universita' degli Studi di Napoli Federico II, Napoli (2009)

Chidichimo, A.: Experimental and analytical investigation on soil-pile-structure interaction. Ph.D thesis, Università Mediterranea Reggio Calabria (2014)

Durante, M.G.: Experimental and numerical assessment of dynamic soil–pile–structure interaction. Ph.D thesis, Universita' degli Studi di Napoli Federico II, Napoli (2015)

Durante, M.G., Di Sarno, L., Taylor, C.A., Mylonakis, G., Simonelli, A.L.: Soil–pile–structure interaction: experimental outcomes from shaking table tests. Earthq. Eng. Struct. Dyn. **45**(7), 1041–1061 (2016). doi:10.1002/eqe.2694

Crewe, A.J., Lings. M.L., Taylor, C.A., Yeung, A.K., Andrighetto, R.: Development of a large flexible shear stack for testing dry sand and simple direct foundations on a shaking table. In: Elnashai, A.S. (ed.) European Seismic Design Practice. Balkema, Rotterdam (1995)

Scasserra, G., Lanzo, G., Stewart, J.P., D'Elia, B.: SISMA (Site of Italian Strong-Motion Accelerograms): a web-database of ground motion recordings for engineering applications. In: Seismic Engineering International Conference Commemorating the 1908 Messina and Reggio Calabria Earthquake, Reggio Calabria, AIP, Melville, NY, vol. 2, pp. 1649–1656 (2008). http://sisma.dsg.uniroma1.it

Durante, M.G., Karamitros, D., Di Sarno, L., Sica, S., Taylor, C.A., Mylonakis, G., Simonelli, A. L.: Characterisation of shear wave velocity profiles of non-uniform bi-layer soil deposits: analytical evaluation and experimental validation. Soil Dyn. Earthq. Eng. **75**, 44–54 (2015). doi:10.1016/j.soildyn.2015.03.010

Helical Screw Piles Performance - A Versatile Efficient Seismic Foundation Systems Alternative for Structures Rehabilitation, New Sustainable Structures Construction and Infrastructure Delivery

Yasser Abdelghany[1(✉)] and Hesham El Naggar[2]

[1] Ministry of Transportation and Infrastructure, University of Victoria, 4D-940, Blanshard Street, Victoria, BC V8W9T5, Canada
Yasser.abdelghany@gov.bc.ca
[2] Western Ontario University, 1151 Richmond Street, London, ON N6A 3K7, Canada
helnaggar@eng.uwo.ca

Abstract. Helical screw anchors have been utilized in uplift forces applications for many years and they have gained popularity for bearing load, seismic performance and dynamic applications and evolved into Helical Screw Piles (HSPs). The speed and ease of their installation, as well as the low cost for new construction and repair, make them versatile for many rehabilitation or new construction applications and as an efficient alternative for deep foundations in infrastructure projects when it is suitable. The main objectives of this paper are to explore the available helical piles systems as well the performance of the helical pile foundation systems under monotonic and cyclic axial and lateral loads. In addition, to explore the new helical screw pile systems suitable for seismic retrofitting of existing foundations and for new structures construction; and to explore their design methodology. Among the helical piles configurations available the circular steel shaft helical piles and the square steel shaft helical piles. Among the different systems the fiber reinforced polymer grouted helical screw piles (FRP-G-HSP); and the reinforced grouted helical screw piles (RG-HSP) in which steel fibers were added to the grout. The research methodology involved conducting more than one hundred full scale field load tests on different instrumented helical piles configurations installed in layered soils profiles. The piles included plain helical screw piles (P-HSP); grouted helical screw piles (G-HSPs); FRP-G-HSPs; and RG-HSPs. The axial and the lateral monotonic and cyclic load tests will be presented as well the piles stiffness and the load transfer mechanisms together with the design theories criteria. The results of 3-D Plaxis finite element modeling were used to establish the load transfer mechanism and to validate the design methodology for the considered piles. Case studies for their applications will be presented.

Keywords: Helical piles · Monotonic · Axial · Lateral · Seismic · Stiffness · Load transfer

T. Abdoun and S. Elfass (eds.), *Soil Dynamics and Soil-Structure Interaction for Resilient Infrastructure*, Sustainable Civil Infrastructures, DOI 10.1007/978-3-319-63543-9_10

1 Introduction and Literature Review

Helical (screw) anchors/piles are available on square shafts (SS) or circular shafts, circular shafts diameters are available up to 36 inches (914.4 mm), larger custom made piles can be designed and installed. The "square shaft" are slender piles that their sections can be enlarged by adding a permanent circular sleeve or my grouting their shafts. SS175 (Square Shaft 1 inch ¾ inch) (Square shaft 44.5 mm*44.5 mm) pile is a segmented deep foundation system with helical steel bearing plates (helices) welded to a central steel shaft, other systems include SS150, SS200 and SS225. Load is transferred from the shaft to the surrounding soil through these bearing plates. Segments or sections are joined with bolted couplings. Practically installed depth is limited only by soil resistance and economics. A helical bearing plate or helix is one pitch of a screw thread. All helices regardless of their diameter have a standard 75 mm pitch for the Square Shaft Piles (Fig. 1). However, the Square Shaft sections pile slender square shaft is susceptible to buckling under loading conditions for such their shafts are encased or grouted to resist buckling. Similarly circular piles as well, the helices have true helical shape and therefore, they do not auger into the soil but rather screw into it with minimal soil disturbance (Fig. 2). The performance of single helical anchors and group action were studied experimentally and theoretically by several researchers, among them Clemence (1983); Mooney et al. (1985) Hoyt (1989), Rao and Prasad (1993), Prasad and Rao (1994).

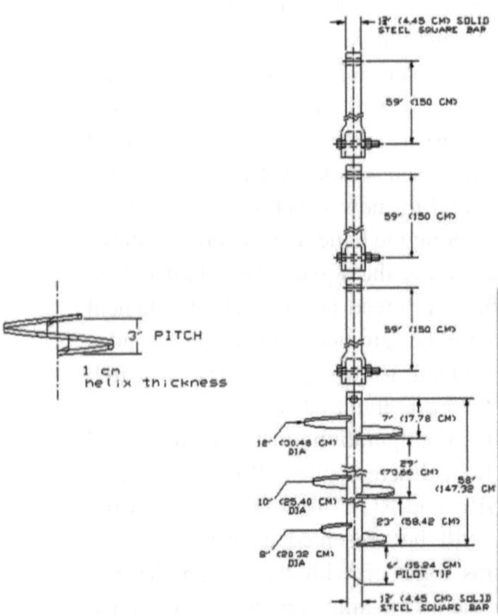

Fig. 1. Schematic of a square shaft SS 175 helical screw foundation system.

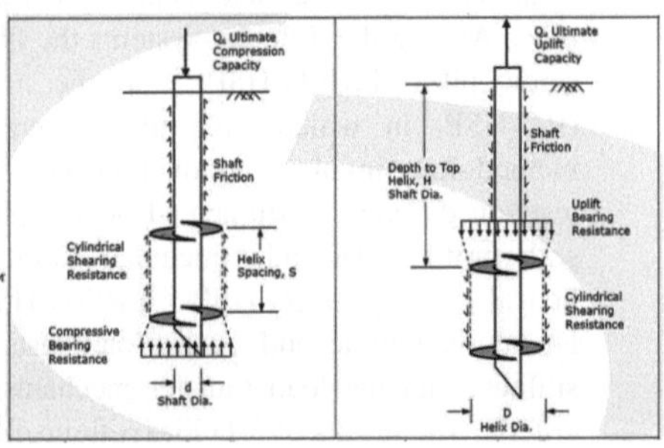

Fig. 2. Schematic of a circular shaft helical screw foundation system under axial/uplift load.

2 Research Objectives and Methodology

I. To develop efficient guidelines for the field installation of (P-HSP) and grouted (G-HSP) and to verify the correlation between the torque of installation and piles axial compression capacities. In addition, investigate the monotonic and cyclic capacities of these piles;
II. Develop new satisfactory helical screw piles systems for seismic zones applications and develop efficient guidelines for their field installation and investigate their performance under monotonic and cyclic loadings;
III. Analyze the load transfer mechanisms and the piles stiffness performance under monotonic and cyclic loading for the existing helical piles and for the Novel ones developed in this research;
IV. Develop a cyclic framework capable of testing helical piles under axial and lateral cyclic loading;
V. Develop a three dimensional (3D) nonlinear, coupled finite element model for helical piles.

The research methodology more than one hundred full scale field load test, in which twenty three piles in which twenty piles were instrumented SS175 helical screw piles systems with three tapered helices 30 cm, 25 cm and 20 cm from top to bottom. Extensions segments of 1.5 m and 2.1 m length were added to the lead section during installation to reach the desired bearing soil stratum, piles were installed up to 5.2 m in layered soil profile.

3 Soil Investigation

Two more boreholes were conducted as part of the research. The two boreholes were advanced to depth 9.6 to 9.8 m by a power auger machine equipped with conventional soil sampling equipment. Standard penetration tests were performed; the results were recorded on the borehole logs as N values according to ASTM D 1586. Shelby tube samples were extracted from both boreholes at equivalent depths to the instrumented installed piles helices in order to properly define the bearing strata and split-spoon samples were also classified. Borehole 1 shows silt and clayey silt overlying stiff to very stiff clayey silt to silty clay layers reaching a very dense fine to medium sand at 8.5 m. Borehole 2 shows silt and clayey silt layers overlying stiff to very stiff silty clay to clayey silt till reaching a very dense fine to medium sand at 9 m approximately. The water table at completion was measured for both boreholes at 5.2 and 6.7 m below the ground surface respectively. Several Triaxial tests were performed in accordance to ASTM (D 2850-95 Re-approved 1999). Table 1 shows sample of the soil properties.

Table 1. Soil properties sample

Borehole and sample depth	Property	Value
Borehole 1-(BH1SA7) depth (3.65–4.25 m) 12–14ft	C_U	40 kPa
	W_C	15.30%
	E	15000 kPa
Borehole 1-(BH1SA9) depth (4.90–5.5 m) 16–18ft	C_U	70 kPa
	W_C	12.00%
	E	45000 kPa
Borehole 2-(BH2SA4) depth 2.15-2.75 m (7–9ft)	C_U	70 kPa
	W_C	12.00%
	E	45000 kPa
Borehole 2-(BH2SA6) depth (3.35–3.95) 11–13ft	C_U	50 kPa
	W_C	17.00%
	E	20000 kPa

4 Grout Testing and Evaluation

Several grouts were evaluated in accordance to find a satisfactory grout to form a grouted column for the helical piles shaft. A series of compression and splitting tensile strength tests were conducted on samples at ages 7 and 28 days. Cylinders groups were prepared using the MS MICROPILE grout, the mean compression strength after 7 and 28 days were 28.6 and 36.3 MPa respectively and their splitting tensile strength were 3.5 and 4.0 MPa respectively. The ASTM C39 and CSA-A 23.13 were followed. Some cylinders were prepared plain (No additives) and the rest were prepared by mixing 1% of NOVOCON 0730 30 mm (1.18 in.) length, 0.7 mm (0.0276 in.) diameter steel fibers to increase their splitting tensile strength. The cylinders prepared with the PT Precision grout without and with steel fibers after the splitting test showed a mean splitting tensile strength of 5.38 and 7.65 MPa after 28 days respectively, while those prepared by Masterflow 1341 showed a mean splitting tensile strength of 4.96 and 5.93 MPa respectively after 28 days. Some cubes of each group were prepared plain (No additives) and others were prepared by mixing 1% of the NOVOCON 0730 steel fibers to study the effect of fibers on the compression strength of the grout. The mean compressive strength for the cubes prepared with the PT Precision grout without steel fibers and others with the steel fibers showed 44.36 and 67.07 MPa after 28 days respectively while the cubes prepared by the Masterflow 1341 grout showed 40.52 and 53.85 MPa respectively.

5 Helical Screw Piles, Instrumentation

The instrumented piles were grouped to plain helical screw piles (P-HSPs), grouted helical screw piles (G-HSPs); grouted reinforced helical screw piles (RG-HSPs), and fiber reinforced polymers grouted helical screw piles (FRP-G-HSPs). The inline torques versus the installation depth of the instrumented piles and for other forty seven plain helical screw piles were also recorded. To determine the axial load distribution along the pile, and more specifically, the load transferred taken by each helix, strain gauges were attached to the shaft of the lead section. Twenty 1.5 m (5 ft) length lead sections were instrumented in which eleven lead sections were instrumented by six strain gauges labeled from one to six, in which strain gauge number one is from the pilot side, near bottom helix, and strain gauge number six is near the top helix. Figure 3 shows a schematic diagram illustrating the strain gauges locations on the lead section shaft. The strain gauges were attached to the shaft very close to the helices (Fig. 4). The remaining nine lead sections were instrumented with eight strain gauges: six strain gauges close to the helices and two strain gauges were installed on the shaft at the mid distance on the shaft between each two helices. Strain gauges #1, #3 and #5 are located below each helix; strain gauges #2, #4, and #6 are located above each helix. The strain gauges labeled as A and B were located in the middle distance between the helices.

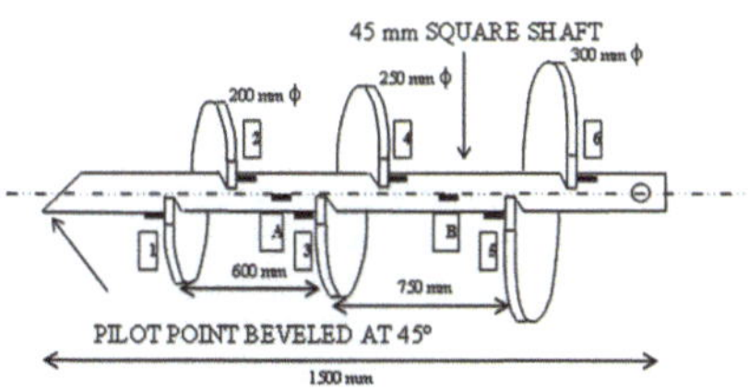

Fig. 3. Schematic lead section instrumentation

Fig. 4. Strain gauges installed

6 Piles Installation and Torqueload Capacity Relationship

The installation torque was recorded for all piles. The capacity of the helical screw pile may be estimated based on the relationship between the installation torque and its ultimate capacity. The principle is that the resistance to installation (defined by installation energy or torque) increases as the helical plates is installed into increasingly stronger soils. Hoyt and Clemence (1989) proposed the following formula for the torque/helical pile capacity relationship:

$$Q_{ult} = K_t T \qquad (1)$$

Where Qult is the ultimate capacity [KN (lb)];

K_t is an empirical torque factor [m^{-1} (ft^{-1})];

and T is the average installation torque along last 1 m of installation (last 3 ft) [kN.m (lb.ft)].

Hoyt and Clemence (1989) recommended $K_t = 33\ m^{-1}$ ($10\ ft^{-1}$) for square shaft HSP of square side dimension smaller than 89 mm. The value of SS systems K_t may range from 10 to 66 m^{-1} depending on soil conditions, shaft size and shape, helix thickness, and application (tension or compression loading).

7 Fiber Reinforced Polymer (Frp)

Fiber reinforced polymer (FRP) tubes of 3 m (10 ft) length and 150 mm (6 in) diameter were utilized to encase eight helical screw instrumented tested piles to provide confinement for the grout and to investigate their seismic performance. The tubes are constructed of continuous glass fibers wound in a matrix of aromatic amine cured epoxy resin in a dual angle pattern that takes optimum advantage of the tensile strength of the filaments. The pipe is manufactured in accordance to ASTM Standard D2996 for filament-wound reinforced thermosetting resin pipe (RTRP). The FRP-G-HSP piles were installed by two different techniques; one in which the grout is provided only inside the tube and the other in which the grout was provide inside and outside the tube to increase the friction component with the adjacent soil. Figures 5 and 6 show respectively FRP-G-HSP Pile assembly and installation.

Fig. 5. FRP-G-HSP Pile prior installation

Fig. 6. FRP-G-HSP installation

8 Monotonic and Cyclic Testing Axial Results

The pile head axial displacement was measured through four HLP 190/FS1/100/4 K linear displacement transducers (LDTs). The displacement average was considered in the data analysis in an attempt to overcome any inaccuracies (Fig. 7). The load cell and LDTs were connected to the data acquisition system. Each instrumented pile was subjected to an initial compression test, followed by a minimum of fifteen cycles of axial loading. A final compression test was conducted after the completion of cyclic loading to examine the piles capacity and performance characteristics during and after cyclic loading. The load transfer mechanism along the pile length was analyzed from the strain gauge records. The spacing between the test and reaction piles complied with ASTM D-1143 and ASTM D-3689. Figure 8 shows axial compression test setup. The cyclic loading was effectively accomplished through a special mechanical-hydraulic setup assembly designed and manufactured as part of this research (Fig. 9).

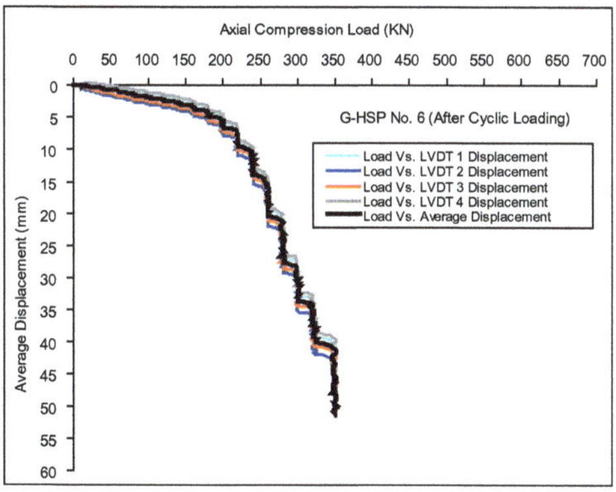

Fig. 7. Load-displacement curves for4 LVDTs and average curve

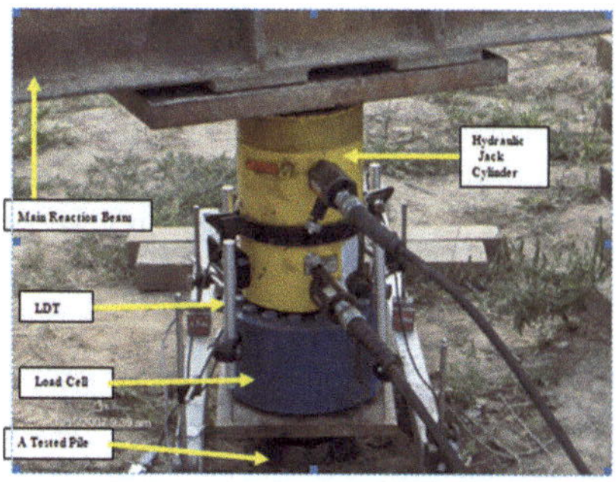

Fig. 8. Axial compression loading

There exist numerous failure criteria that are used for different pile types and in different building codes. The one suggested by Terzaghi (1940) was used; for practical purposes, the ultimate load should be defined as that which causes a settlement of one-tenth of the pile diameter or width. The failure criteria place the ultimate load within the nonlinear region of the load-movement curve to ensure that once a suitable factor of safety is applied, the design load of the pile should lie within the initial linear region of the curve. This will yield predictable load-displacement behavior and avoid any abrupt settlement. The axial pile load tests were conducted according to the ASTM D-1143 standard test method for piles under static axial compression load and under axial cyclic load. The quick testing method has been used successfully to test helical piles.

Based on the axial load testing results analyses, the following can be drawn:

(1) The piles capacities were found to be proportional to the installation torque. Therefore, the empirical torque correlation factor K_t can be used to predict the pile capacity of the plain helical screw piles (P-HSP). In axial compression, the value of K_t of 33 m−1 is a sound value for piles in clayey silt to silty clay soils. Samples of the axial results are presented in (Figs. 10, 11 and 12).

(2) The Terzaghi (1940) failure criterion (10% of the average helices diameter) was adopted to obtain the ultimate axial compression capacities of all tested piles. It was found that the capacity of piles before cyclic loading varied between 240–282 KN for P-HSPs, 321–341 KN for G-HSPs, 235–327 KN for FRP-G-HSPs with internal grout, 303–460 KN for FRP-G-HSP piles of internal and external grout, and 431–650 KN for RG-HSP piles.

(3) The capacity of piles after 15 load cycles varied between 278–313 KN for P-HSPs, 280–422 KN for G-HSPs, 264–483 KN for FRP-G-HSPs with internal grout, 290–338 kN for FRP-G-HSPs of internal and external grout, and 553–617 kN for RG-HSPs.
(4) Minimal degradation of piles stiffness occurred after the 15 loading cycles, where the reinforced grouted helical screw piles (RG-HSP) presented the best stiffness performance.
(5) The reinforced grouted helical screw piles (RG-HSP) showed the highest axial ultimate compression capacity of all different geometry tested helical piles. This confirms the beneficial effect of the reinforced grouted shaft on increasing the axial capacity and enhancing the seismic performance.
(6) The load transfer mechanism analyzed from the measured strain data showed about 55% shaft resistance in case of the reinforced grouted helical screw piles, and an average of 14% in case of plain helical screw piles (Fig. 13).

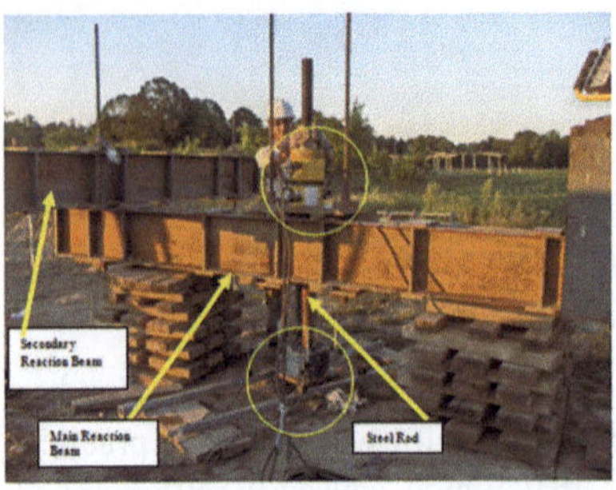

Fig. 9. Axial cyclic loading setup

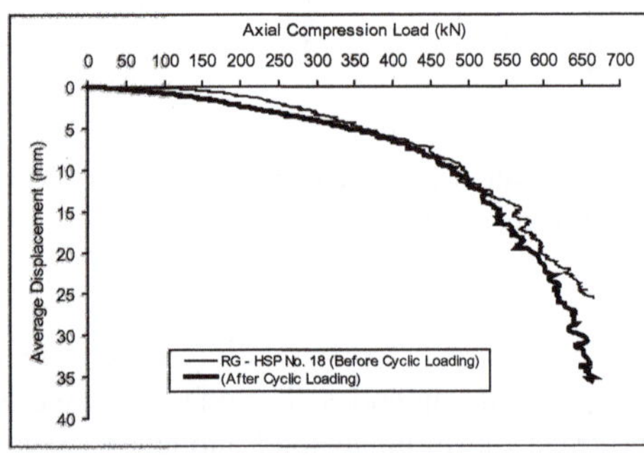

Fig. 10. Load-displacement RG-HSP 18 before and after cyclic loading

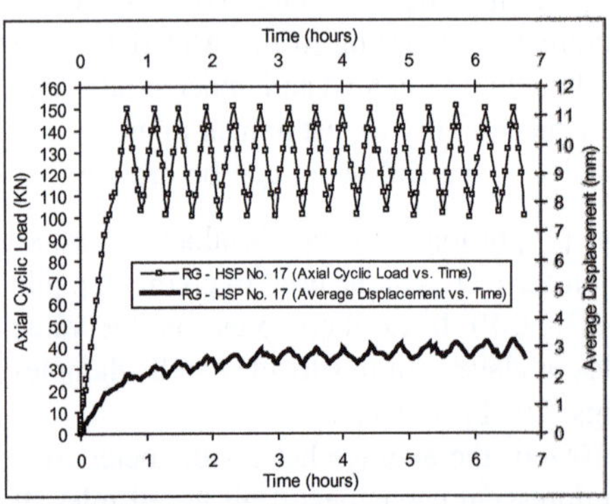

Fig. 11. RG-HSP 17 axial cyclic load-displacement

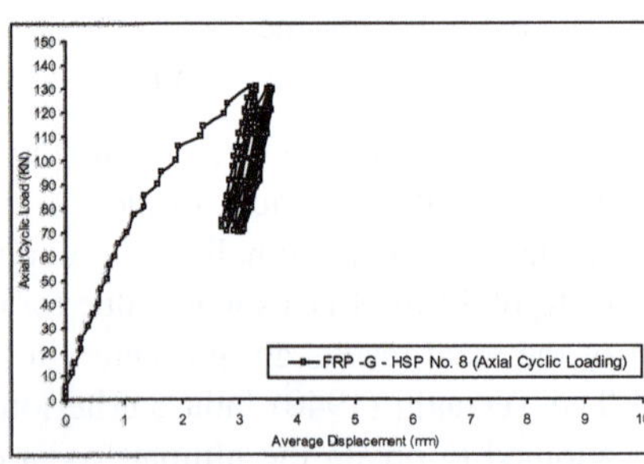

Fig. 12. FRP-G-HSP 8 axial cyclic load-displacement

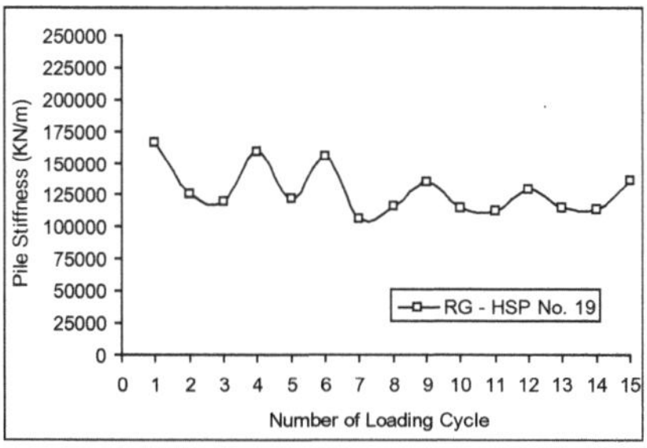

Fig. 13. RG-HSP 19 pile stiffness versus loading cycles

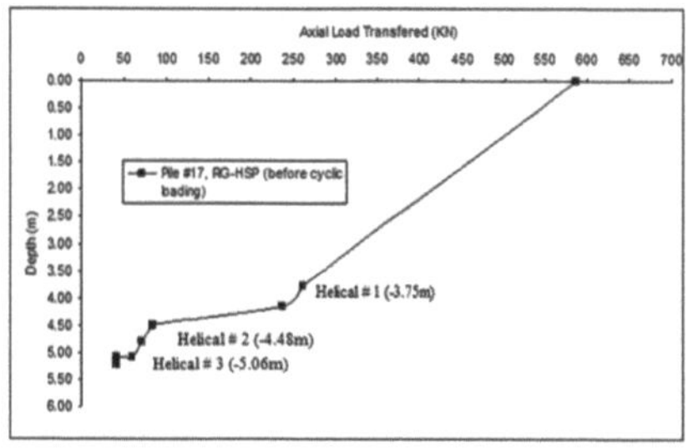

Fig. 14. RG-HSP 17 axial load transfer

9 Monotonic and Cyclic Lateral Testing Results

Twenty piles were subjected to lateral loading. An initial lateral load test was performed on each pile, followed by fifteen cycles of lateral loading. After the completion of cyclic loading, each pile was subjected to a monotonic lateral load test to determine the pile lateral capacity after cyclic loading. The pile lateral load-displacement curve can be used to evaluate the pile's performance under lateral loading and to assess its ultimate capacity. A generally accepted ultimate lateral load criterion is defined as the load that corresponds to a lateral displacement at the pile head equal to 6.25 mm (Prakash and Sharma 1990) (Fig. 14).

Based on the lateral load testing results analyses, the following can be drawn:

(1.) The ultimate capacities of the tested piles were obtained as the load at pile head deflection of 6.25 and 12.5 mm (i.e. two different failure criteria). The P-HSPs had negligible lateral capacity. The capacity of the G-HSPs varied between 14 and 26 KN and from 7.5 to 12 KN for FRP-G-HSPs with internal grout and from 20 to 64 KN for FRP-G-HSPs with internal and external grout. The RG-HSPs showed the highest lateral capacity, which ranged from 42 to 80 kN. Similar trends (but higher values) were observed for the capacity based on 12.5 mm deflection. (2.) The lateral capacity of most pile configurations showed some degradation due to the cyclic loading. However, the RG-HSPs showed a small reduction, and in some cases some increase, in the capacity after the cyclic loading. (3.) The reinforced grouted helical screw piles (RG-HSP) presented the best stiffness performance during the 15 loading cycles. Samples of the lateral results are presented in Figs. 15, 16, 17 and 18.

Fig. 15. FRP-G-HSP (external grout) After cyclic lateral testing

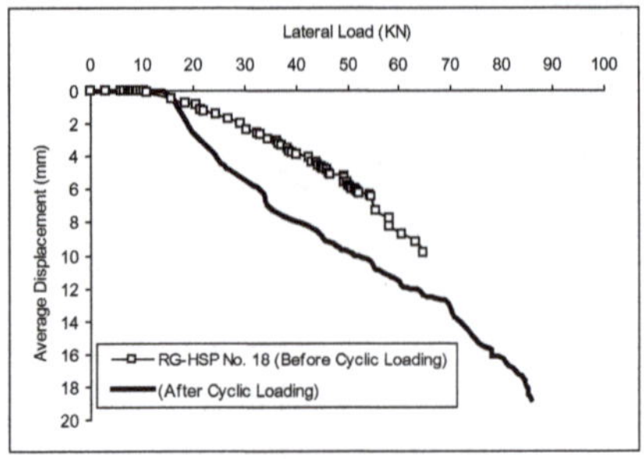

Fig. 16. RG-HSP lateral load-displacement (before and after cyclic)

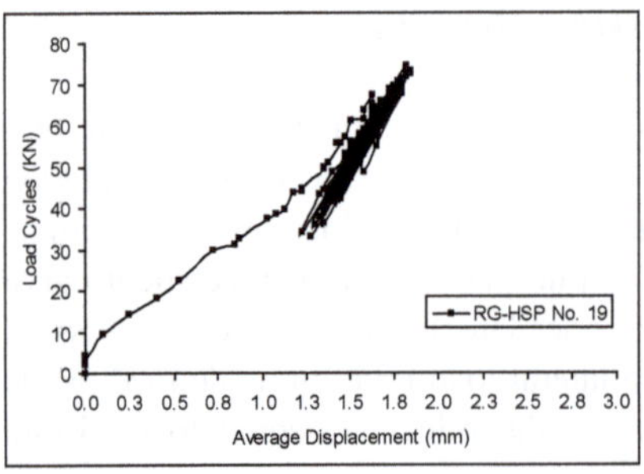

Fig. 17. RG-HSP lateral cyclic load-test

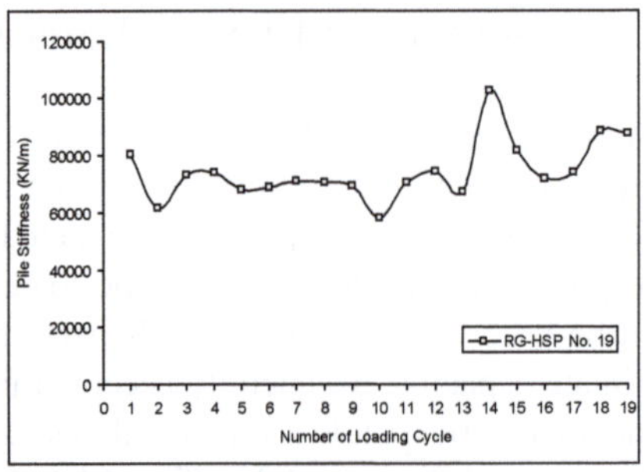

Fig. 18. Stage 3 – RG-HSP stiffness versus number of loading cycles

10 Helical Screw Piles Finite Element Modeling Plaxis 3d

In principle, Plaxis 3D provides an advanced deformation-based tool to analyze pile and raft foundations, offshore foundations and similar problems in the geotechnical field (Brinkgreve and Swolfs 2007). A two-dimensional mesh of the geometry is created in the software environment by means of work-planes. This step is followed by the generation of a three dimensional mesh. PLAXIS automatically imposes a set of generated fixities to the boundaries of the model. A typical distribution of elements and their boundary conditions are illustrated in Fig. 19 and Table 2. Sample comparisons between the measured and simulated Model results are highlighted in Figs. 20, 21 and 22. A finite element model was developed using PLAXIS 3D Foundation software. Based on the available soil properties from the field testing analyzed data, the model was adjusted to best emulate the full-scale testing conditions. The model was provided for the base line plain helical screw piles (P-HSP), the reinforced grouted helical screw piles (RG-HSP) - which provided the highest axial and lateral capacities - and for the fibre reinforced polymer helical screw piles (FRP-G-HSP). Based on the comparison between the finite element models developed for the base line piles SS175 plain helical screw piles (P-HSP), the reinforced grouted helical screw piles (RG-HSP), and the fibre reinforced polymer grouted helical screw piles (FRP-G-HSP), the numerical and field test results showed favourable match under axial and lateral loadings. The results from the numerical model were used to verify the load transfer mechanism for RG-HSPs (Fig. 23).

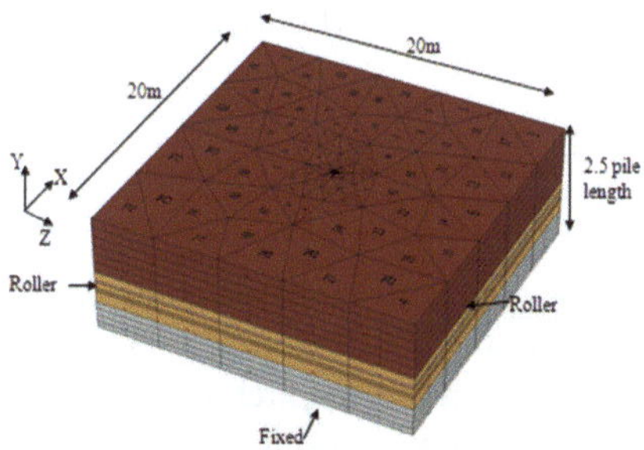

Fig. 19. Finite element mesh RG-HSP 17 ft (5.2 m) length

Table 2. Boundary conditions for modeled piles

Pile	Description	length m (ft)	shaft cm^2	Model boundaries width m^2	Model boundaries Depth m
P-HSP (stage #1) Axial loading	Plain helical screw pile	3.6 (12)	Square, 4.5x4.5	20x20	8.5
P-HSP (stage #3) Lateral loading	Plain helical screw pile	5.2 (17)	Square, 4.5x4.5	20x20	9.0
RG-HSP (stage #3) Axial & lateral loading	Reinforced grouted helical screw pile	5.2 (17)	Diameter = 15	20x20	9.0
FRP-G-HSP (stage #3) Lateral loading	Fiber reinforced polymer grouted helical screw pile	5.2 (17)	Diameter = 15	20x20	9.0

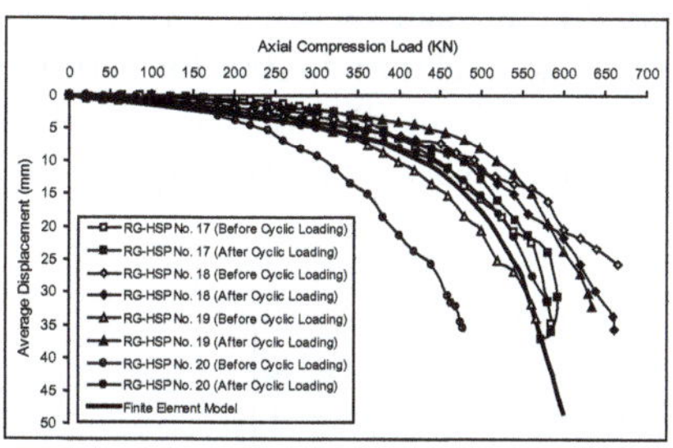

Fig. 20. Measured and simulated axial load-settlement (RG-HSP) 5.2 m (17 ft)

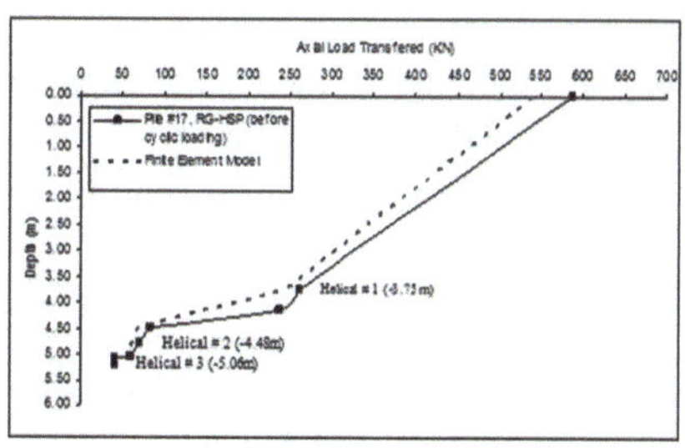

Fig. 21. Axial load transfer measured and simulated distribution in (RG-HSP)

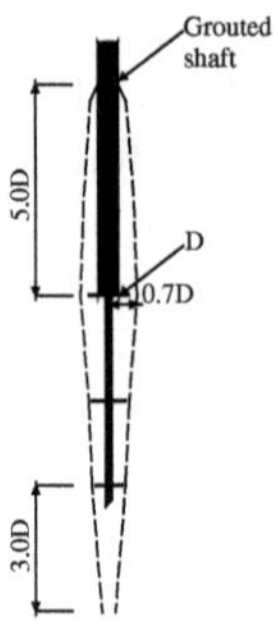

Fig. 22. Surface that traces the points reached shear failure for Axial RG-HSP

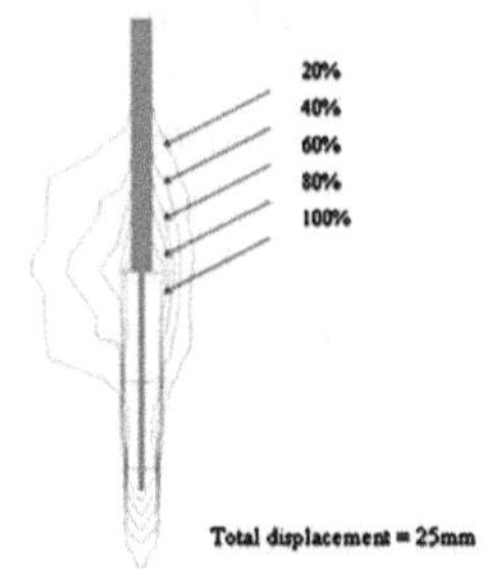

Fig. 23. Soil total displacement contours for RG-HSP at ultimate load

11 Summary and Conclusions

The monotonic and cyclic performance of helical piles under axial and lateral loading conditions was evaluated through full-scale field load testing. The piles include SS175 (P-HSP), (G-HSP), the novel (FRP-G-HSP) developed from this research in which the helical piles shafts were installed encased in FRP tubes without causing damage to the FRP tubes during installation; The other Novel pile is the (RG-HSP), in which steel fibers were mixed to the satisfactory grout to increase its tensile splitting strength. The P-HSPs axial capacities agreed well with empirical formula relating the installation torque to the pile capacity with torque correlation factor, Kt = 33 m−1 for compression loading. The cyclic performance of P-HSPs and G-HSPs were satisfactory and warranted a consideration for seismic applications. P-HSPs and G-HSPs experienced minimal reduction in their axial capacities after 15 load cycles of 5–10% for P-HSPs while the G-HSPs showed a variation of about ±18%. Both the (P-HSPs and G-HSPs) lateral capacities were low due to the slender shaft. The axial capacity of the

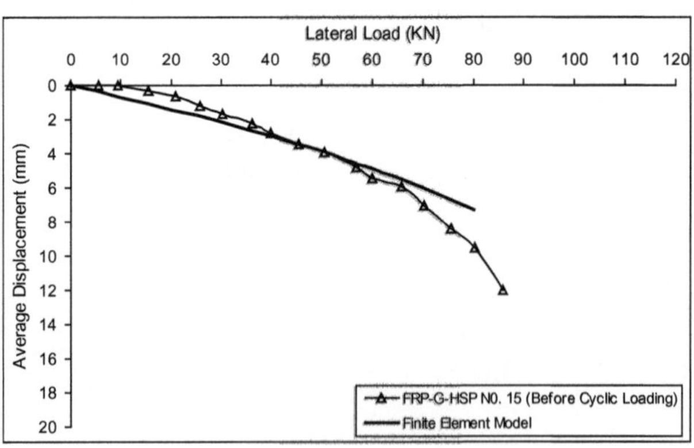

Fig. 24. (FRP-G-HSP) Lateral load-deflection Measured and simulated

(FRP-G-HSPs) was slightly higher than the P-HSPs but the lateral capacity and stiffness of the internally and externally grouted (FRP-G-HSPs) were twice of the (FRP-G-HSPs) with internal grout only and 3 times more than the (P-HSPs). The (RG-HSP) showed the best performance among all tested piles under axial and lateral Monotonic and cyclic loadings in which the (RG-HSPs) axial capacity was more than twice that for (P-HSP), with minimal reduction after cyclic loading, and their lateral capacity was more than 3 times the (P-HSPs) capacity. A 3-D finite element model using Plaxis software was created to develop a design methodology for the considered piles (Fig. 24).

Acknowlegements. The authors want to thank NSERC, OCE (Ontario Center of Excellence), Hubbell-Chance, BASF the chemical company, Pipes Specialties, and Propex Concrete systems for supporting this research.

References

Abdelghany, Y., El-Naggar, M. H.: Full-Scale Experimental Investigations & Numerical Analysis of Different Innovative Instrumented Composite Helical Screw Piles under Axial & Lateral Monotonic & Cyclic Loadings. In: Vancouver Geotechnical Society Symposium on Foundation and Lifeline Engineering, 7th June, 2013, Vancouver, British Columbia (2013a)

Abdelghany, Y., El-Naggar, M.H.: Innovative seismic resistant helical screw piles. In: CSCE Conference, 29th May–1st June, 2013, Montreal, Quebec (2013b)

Abdelghany, Y., El-Naggar, M.H.: Performance of instrumented composite helical screw piles seismic systems in layered soils. In: 65th Canadian Geotechnical Conference, 30th September–3 October, 2012, Winnipeg, Manitoba (2012)

Abdelghany, Y., El-Naggar, M.H.: Steel fibers reinforced grouted and fiber reinforced polymer grouted screw piles – an innovative deep foundations seismic systems. In: 64th Canadian Geotechnical Conference, October 2–6, 2011, Toronto, Ontario (2011b)

Abdelghany, Y., El-Naggar, M.H.: Monotonic and cyclic performance of helical screw piles under axial and lateral loadings. In: Fifth International Conference on Recent Earthquakes Engineering and Soil Dynamics, May 24–29, 2010, San Diego, CA (2010a)

Abdelghany, Y., El-Naggar, M.H.: Full-scale experimental investigations & numerical analysis of different innovative instrumented helical screw piles under axial & lateral monotonic & cyclic loadings. In: 63th Canadian Geotechnical Conference, Calgary, Alberta, Canada, 12–15, September 2010 (2010b)

Abdelghany, Y., El-Naggar, M.H.: Full-scale experimental and numerical analysis of instrumented helical screw piles under axial and lateral monotonic and cyclic loadings – a promising solution for seismic retrofitting. In: ASCE IECC6 6th International Engineering and Construction Conference, June 28–30, 2010, Cairo, Egypt (2010c)

Abdelghany, Y.: Monotonic and cyclic behavior of helical screw piles under axial and lateral loading. Ph.D. thesis, The University of Western Ontario, Canada (2008)

ASTM International. Standard Test Method for Individual Piles under Static Axial Tensile Load. Designation D3689-90 (Reapproved 1995). West Conshohocken, PA, USA

ASTM International. Standard Test Method for Individual Piles under Static Axial Compressive Load. Designation D 1143–81 (Reapproved 1994). West Conshohocken, PA, USA

Brinkgreve, R.B.J., Swolfs, W.M.: PLAXIS 3D Foundation. Version 2 Manual (2007)

Clemence, S.P.: Dynamic uplift capacity of helical anchors in sand. Civil Eng. Pract. Des. Eng. **2** (3), 345–367 (1983)

El-Naggar, M.H., Abdelghany, Y., El Sharnouby, M., Frater, R.: Seismic restraint helical pile systems and method and apparatus for forming the same, US PATENT 27 December, 2012. (US2012/0328374-A1)

El-Naggar, M.H., Abdelghany, Y.: Seismic helical screw foundations systems. In: Proceedings of the 60th Canadian Geotechnical Conference, Ottawa, Ontario, Canada, 21–24, October 2007 (2007a)

El-Naggar, M.H., Abdelghany, Y.: Helical Screw Piles (HSP) capacity for axial cyclic loadings in cohesive soils. In: Proceedings of the 4th International Conference on Earthquake Geotechnical Engineering, Thessaloniki-Greece in 25–28, June 2007 (2007b)

Hoyt, R.M., Clemence, S.P.: Uplift capacity of helical anchors in soil. In: Proceedings of the 12th International Conference on Soil Mechanics and Foundation Engineering. Rio de Janerio, vol. 2, pp. 1019–1022 (1989)

Mooney, J.S., Adamczak, S.J., Clemence, S.P.: Uplift capacity of helical anchors in clay and silt. In: ASCE Convention Conference Proceedings, Detroit, MI, pp. 48–72 (1985)

Prakash, S., Sharma, H.D.: Pile Foundations in Engineering Practice. Wiley, New York (1990)

Prasad, Y.V.S.N., Rao, N.S.: Pullout behaviour of model pile and helical pile anchors subjected to lateral cyclic loading. Can. Geotech. J. **31**(1), 110–119 (1994)

Rao, N.S., Prasad, Y.V.S.N.: Estimation of uplift capacity of helical anchors in clay. J. Geotech. Eng. **119**(2), 352–357 (1993)

Terzaghi, K.: Theoretical Soil Mechanics. Wiley, New York (1940)

Laterally Loaded Test for Pile with Upper Soil Grouted

Guangming Yu, Weiming Gong(✉), Guoliang Dai, and Meihe Chen

Key Laboratory for RC and PRC Structures of Education Ministry, Southeast University, Nanjing 210096, China
yuguang_ming@163.com, {wmgong,daigl}@seu.edu.cn, 597201998@qq.com

Abstract. Laterally loaded piles may entail large deformation under some circumstances such as ship impact, earthquake and long-term horizontal thrust of arch bridges, etc. The arch bridges are especially sensitive to horizontal displacement of foundation. Lateral load test was performed on a full scale pile in virgin soil and pile with post-grouted soil around pile. Further numerical simulation analysis was carried out by PLAXIS3D based on field test result. The analytical and measured results are found to be in fairly satisfactory agreement. According to test results of single pile, lateral response of piles largely depends on profile of limiting force in shallow soil. Horizontal bearing capacity of bored piles is improved via grouting technique. For pile with post-grouted soil around, horizontal displacement of pile top is significantly reduced. The maximum bending moments of pile with grouting decreases dramatically. The grouted upper soil expands range of effective soil around pile, more horizontal thrust was shared.

Keywords: Thrustful foundation · Lateral displacement · Upper soil grouted · Load sharing

1 Introduction

At present, drilled grouting pile remain the preferred foundation scheme for super high-rise buildings and large-scale bridges, which have many advantages of strong bearing capacity of single pile, little settlement and deformation easily controlled and so on. At the same time, post grouting technique has been widely applied to drilled grouting pile foundation engineering construction etc. This technique has been used in engineering practice for many years to develop more perfect construction technology and method than before. A more profound understanding of its functional mechanism is acquired by engineering personnel. The cement slurry is pressed into the mud around pile by upper soil post-grouted technique. That contributes to developing coarser interface between pile and soil and increase friction between them.

In reinforced concrete facility such as Highway Bridge, port and pier and offshore oil platform etc. Horizontal load is applied in these infrastructures more frequently. Even combined load applied which include lateral load, vertical load and bending moment. According to a large number of literatures (Guo (2004a), Poulos (1971), Randolph and Houlsby (1984, 1988), Wan (2015)), response of pile under lateral load

T. Abdoun and S. Elfass (eds.), *Soil Dynamics and Soil-Structure Interaction for Resilient Infrastructure*, Sustainable Civil Infrastructures, DOI 10.1007/978-3-319-63543-9_11

is mainly in the range of upper soils around piles. So this measurement that it is post-grouted only in the upper soil around piles can satisfy the requirement of practical engineering. There are significantly applied values to study horizontal load-bearing properties of piles with soil post-grouted around.

2 Project Profile

New street across the Luo River bridge connects north with south of Luo River in Luoyang city. It consists of south, north and river channel section of the bridge. Main bridge is designed for arch bridge in river channel section and its topside structure for nine spans arch with a single arch span of 45 m–95 m. Substructure is designed for low-cap grouting pile foundation which adopts rotary digging bored piles construction technology. Bridge floor width of 35.5 m, total length of 618 m, the north approach span length of 289 m, the south approach length of 206 m, the flood control standard for 100 a.

Arch bridge is a kind of statically indeterminate structure bearing long-termly horizontal thrust. It is very sensitive to foundation's displacement, especially lateral displacement. In some circumstance pile group foundation is adopted, subgrade just provides part constraints to horizontal displacement. But lateral displacement of piers is likely to exceed the allowed limit and change arch axis of main arch ring in adjacent cross when arch bridge produce long-term lateral force to piers. As a result, large additional bending moment is produced in the main arch ring so that the adjacent cross collapsed. Even bridge collapsed in a row.

2.1 Geological Conditions and Soil Parameters

There are 9 layers of soil around piles. Soil parameters which show in Table 1 are used to calculate the response of piles in PLXIS3D. Figure 1 shows particle size distribution of sand contained in gravel. By analyzing the sieve analysis test data, it was found that the gradation of sand was bad, with even coarse grains with diameters mostly exceeded 0.5 mm.

Table 1. Soil parameters

Item	Natural unit weight (KN/m3)	Friction resistance (kPa)	Compression modulus (MPa)	Internal friction angle (°)
⑥-2Fine sand	19.2	35	15	23
⑥-3Round gravel	20	70	22	27
⑥Gravel	21.2	100	30	33
⑨-2Round gravel	20.8	90	27	30
⑨Gravel	21.6	140	35	35
⑩-2Round gravel	21	110	30	32
⑩Gravel	22	170	40	38

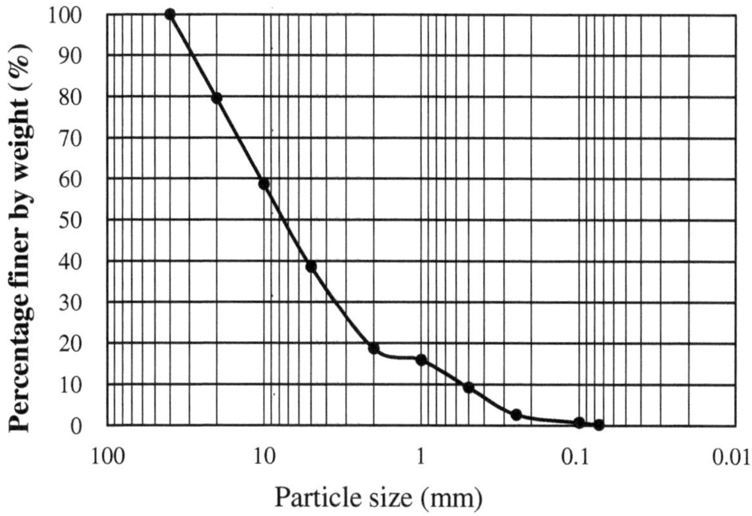

Fig. 1. Particle size distribution of sand

2.2 Grouting Scheme

Pile under vertical load ability is stronger, but a lot of engineering pile top will bear horizontal load, Piles' vertical bearing capacity is higher than horizontal bearing capacity. In a vast quantity of engineering, however, pile foundations have to bear lateral load. Soil

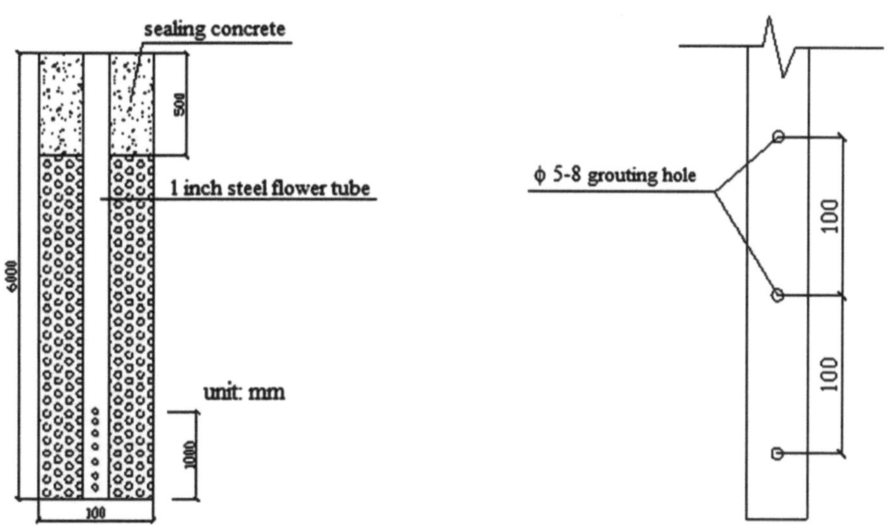

(a) Structure diagram of grouting hole **(b) Structure diagram of steel flower tube**

Fig. 2. Grouting scheme

post-grouted technique was used in drilled grouting pile foundation engineering construction in order to improve its bearing capacity so that safety of arch bridge can be ensured. The piles have a diameter of D = 2.0 m and a length of L = 35.0 m (relative length L/D = 17.5). Figure 2 shows grouting scheme. The Chinese JGJ design code (JGJ94-2008) recommends mainly influenced depth under lateral load is 2(D + 1), so depth of grouted soil is 6 m. Grouting pipe's diffusion radius is 0.5 m. The range of grouted soil around pile is half ring with 1.0 m in diameter away from lateral sides of piles.

3 Establishment of Numerical Model

We performed numerical simulation of lateral load tests process employing the FE method, using the geotechnical finite element analysis software PLAXIS3D. Two pile models was calculated, general drilled grouting pile and drilled grouting pile with upper soil grouted, respectively. The soil model range is 30 m*30 m, and 60 m in depth, which is consisted of 9 layers of soil. Volume pile from PLAXIS is chosen to simulate mechanical characteristic of pile. We modelled soil with tetrahedral continuum finite elements. We employed the elasoplastic constitutive relationship: Mohr-coulomb failure criterion. In this test, lateral surface load is applied at the top of pile according to slowly maintaining load method. It is divided into 9 levels of load with increment of 150 KN. The maximum load value is 1500 KN. Figure 3 shows the distribution of soil layers.

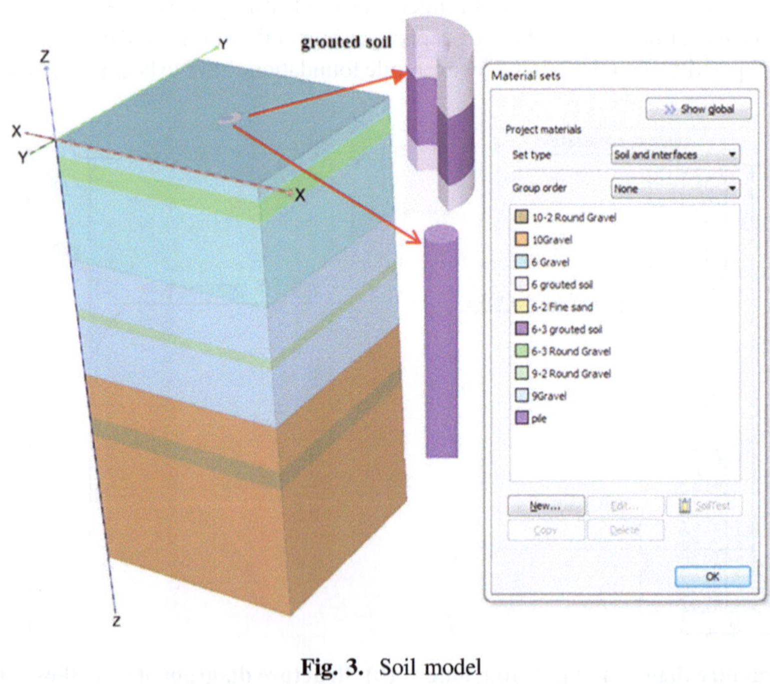

Fig. 3. Soil model

4 Parameters Back Analysis

Figure 4 shows piles' lateral displacements in loading and unloading process for pile in virgin soil and pile with upper soil grouted. In this test process, lateral loads vary from 300 KN to 1500 KN at the increment of 150 KN, and load divided into 9 levels, unload divided into 5 levels. It can be concluded from Figure that the finite element analytical and measured results are found to be in fairly satisfactory agreement. Therefore, the further analysis of pile-soil interaction is conducted via PLAXIS on the basis of testing data, which makes up for the deficiency that the particular situation of piles-soil interaction cannot be observed under the existing test condition. Residual deformation of pile head is 4.2 mm for pile in virgin soil after unloading, while it is only 2.5 mm for pile with upper soil grouted. By parameters back analysis method, it is suitable that Mohr coulomb model is used for grouted soil. Its Young's modulus is 3.4 times than virgin soil, and can increase from 0.5 kPa to 19, internal friction angle can be changed from 27° to 40°.

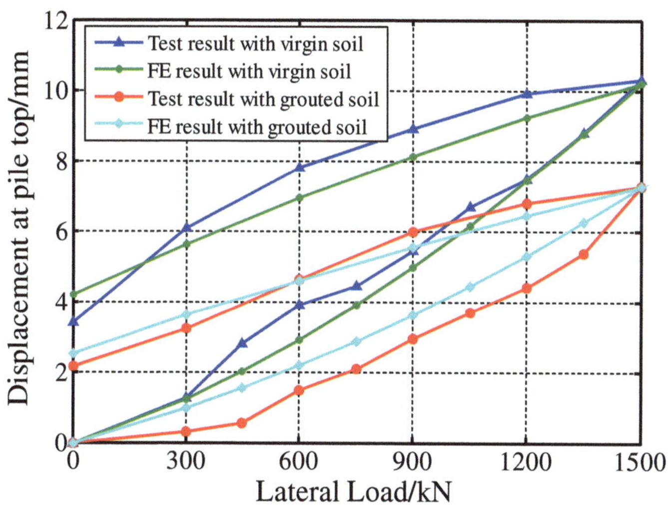

Fig. 4. Lateral load vs. displacement curve

5 Analysis of Results

5.1 Displacement of Piles

Figure 5 shows deformation curve of pile shaft from PLAXIS result for two kinds of pile under different levels of horizontal load. In loading process, deformation of pile head varies from 1.23 mm to 10.2 mm for pile in virgin soil, while it varies from 0.97 mm to 7.27 mm for pile with upper soil grouted under the same levels of lateral load, and increases slowly. The depth of the lateral displacement response is −10.5 m when load reaches to 1500 KN. It is −10.2 m for pile with upper soil grouted. So there is no distinct difference in the depth of the lateral displacement response. Therefore, it is soil grouted around pile that enhances the ability of resistance to lateral

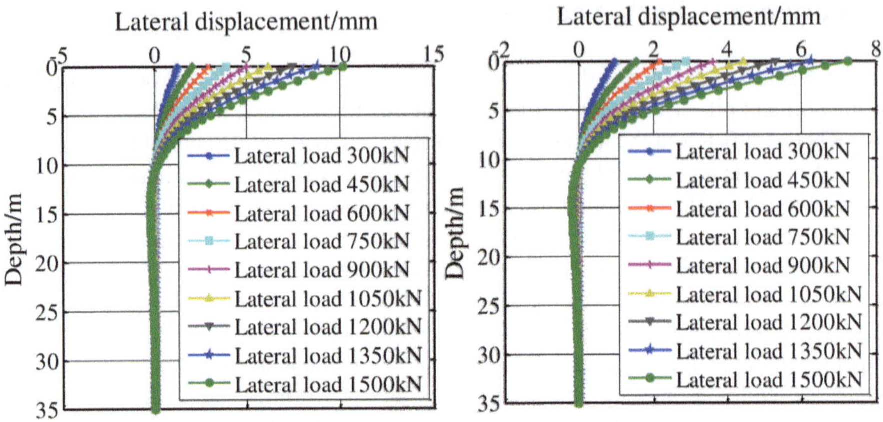

(a) pile in virgin soil **(b) pile with grouted soil**

Fig. 5. Pile deformation vs. depth curve

displacement of pile. This method of soil post-grouted around pile can be considered to apply to pile foundation engineering, especially when it is sensitive to lateral displacement, improving the safety and reliability in engineering.

5.2 Moments of Piles

Figure 6 shows piles bending moment distribution of pile in virgin soil and pile with upper soil grouted under different levels of horizontal load. The bending moments of pile in virgin soil increase more quickly than pile with grouting. Maximum moments of pile in virgin soil varies from 460 KN•m to 3437 KN•m as different levels of lateral

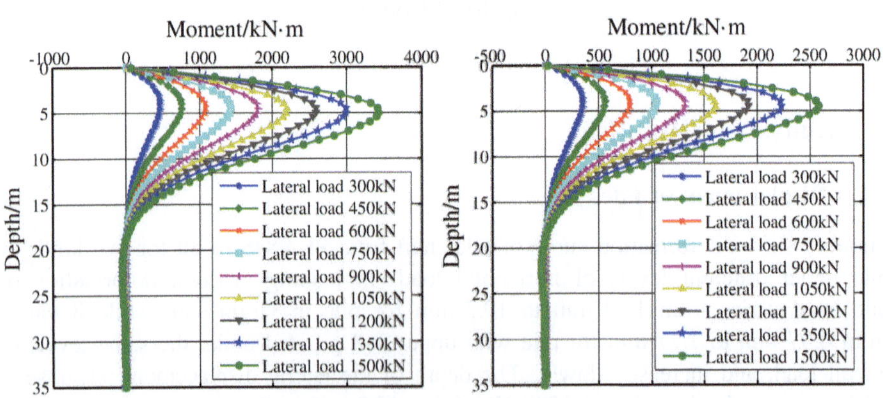

(a) pile in virgin soil **(b) pile with grouted soil**

Fig. 6. Bending moments vs. depth curve

load is applied. However, Maximum moments of pile with grouting varies from 346 KN•m to 2580 KN•m as different levels of lateral load is applied. The Maximum moments decreases by 25%.

Therefore, this treatment method that soil around pile is post-grouted can make moments of piles decrease dramatically and improve the safety of pile greatly. More horizontal thrusts are shared by soil around pile.

5.3 Shear Force of Pile

Figure 7 shows pile's shear force distribution of pile in virgin soil and pile with upper soil grouted under different levels of horizontal load. Pile's maximum shear force locates in the top of pile. From the perspective of shear overall distribution, shear force of pile with grouting along depth is less than pile in virgin soil. The maximum shear force of lower part of pile in virgin soil varies from 54 KN to 433 KN·m as different levels of lateral load is applied, while the maximum shear force of pile with upper soil grouted in the same position varies from 39 KN to 323 KN with increasing lateral load, it reduces by 25% under maximum load value.

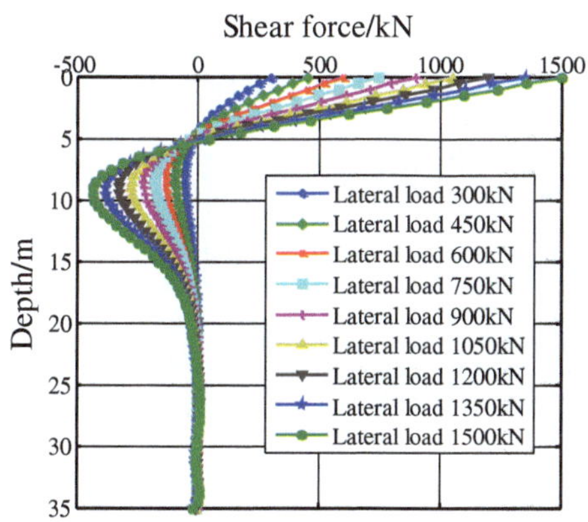

(a) pile in virgin soil

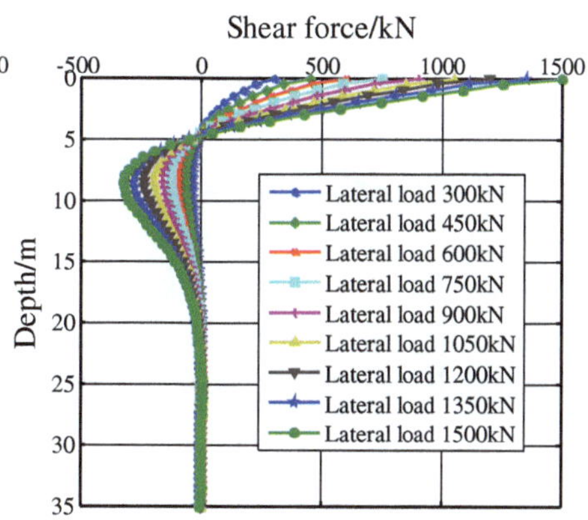

(b) pile with grouted soil

Fig. 7. Shear force vs. depth curve

5.4 Nephogram Analysis

5.4.1 Planar Nephogram Analysis

Figure 8 shows lateral displacement nephogram of soil around pile. Initial phase in Fig. 8 (a) is K0 procedure that initial effective stresses, pore pressures and state parameters are generated directly. There is no displacement in this calculation.

Figure 8 (b)–(d) show different lateral displacement planar nephogram at mud surface when loads are applied at top of pile from 300 KN, 750 KN and 1500 KN. Compared with pile in virgin soil, influence scope of soil subjected lateral load is larger for pile with upper soil grouted and has smaller deformation. Along the direction of

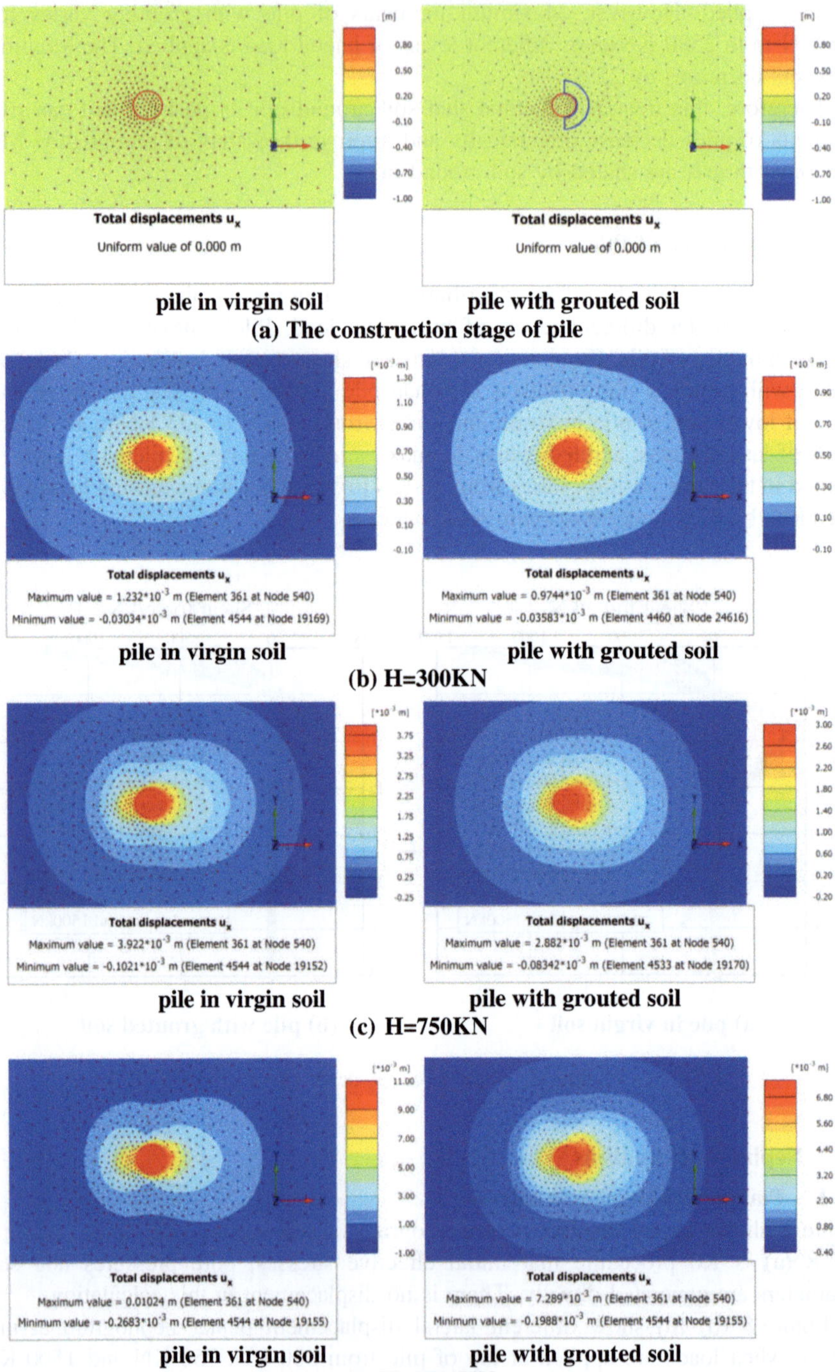

Fig. 8. Planar nephogram of soil

loading, displacement isoline of pile with grouting is sparser than pile in virgin soil's. Along the direction perpendicular to the loading, effective soil resisting lateral load expands dramatically for pile with grouting. The range of treated soil increases significantly in terms of lateral displcement.

5.4.2 Profile Nephogram Analysis

Figure 9 shows lateral displacement profile nephogram of soil around pile under 1500 KN. In terms of working soil profile, influence scope of soil subjected lateral load is also larger for pile with upper soil grouted than for pile in virgin soil. Expanding soil shares more lateral loads so that horizontal displacement of pile with grouting becomes smaller. It can be concluded that increasing strength of grouted soil results in decreasing displacement field from Fig. 10.

Ashour et al. (1998, 2013) used an envisaged 3D strain wedge (SW) model to determine pile reaction under lateral loads. The SW model was able to link between the more complex 3D soil-pile interaction and a simpler one-dimensional characterization. Figure 11 shows shape of soil resistance on front of pile is like three dimension wedge.

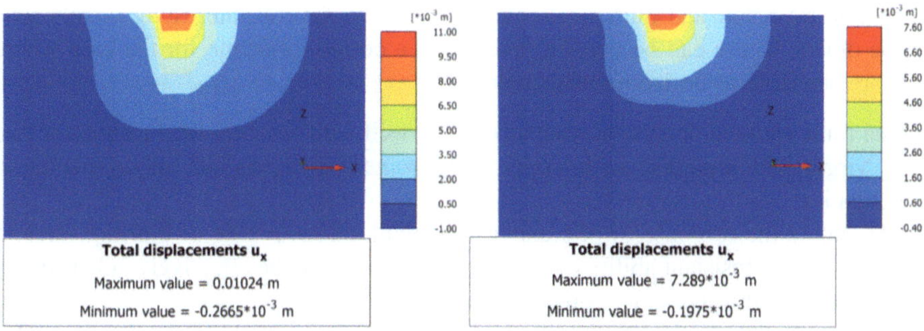

pile in virgin soil **pile with grouted soil**

Fig. 9. Profile nephogram (1500 KN)

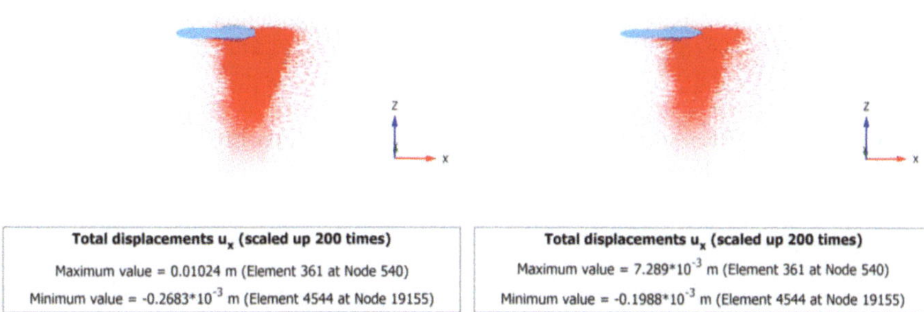

pile in virgin soil **pile with grouted soil**

Fig. 10. Displacement field of pile and soil (1500 KN)

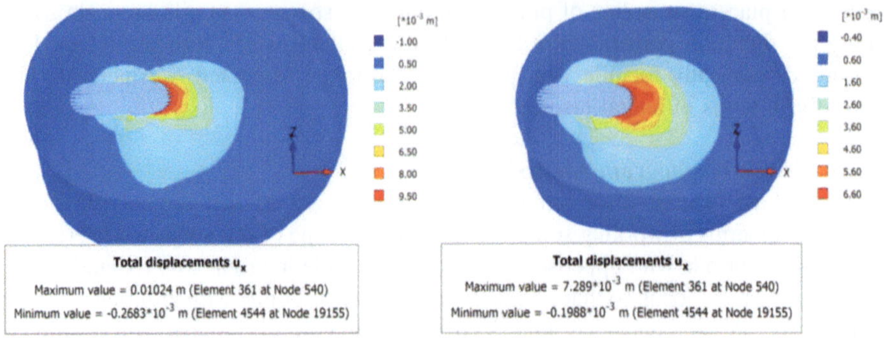

pile in virgin soil **pile with grouted soil**

Fig. 11. ISO surfaces of lateral displacement (1500 KN)

6 Conclusion

From this test result and numerical calculation study, several interesting conclusions can be drawn. These can be used in practice as a design guideline for the case where piles are sensitive to horizontal displacement.

1. The finite element analytical results coincide well with the measured results for pile in virgin soil and pile with upper soil grouted. It meets the requirements of validation.
2. The key to the response of pile under horizontal load is profile of limiting force in shallow soil. Resistant ability of lateral displacement can be effectively enhanced by this method of upper soil grouted around pile.
3. Compared with un-grouted pile, the value of maximum bending moments for grouted pile and pile's shear force reduce with superior safety of pile.
4. Soil around pile is treated with post-grouted technique, the range of effective soil providing resistance expands to share more lateral load, reducing displacement of pile head.
5. In terms of reducing lateral displacement, this treatment method that soil around pile is post-grouted can be considered to adopt in pile foundation sensitive to lateral displacement

Acknowledgments. The research work presented in this paper is supported by the National Basic Research Program of China (973 Program, Grant 2013CB036304) and the National Natural Science Foundation of China (NSFC) (Grant 51478109), for which the authors are grateful.

References

Guo, W.D.: On limiting force profile, critical depth, and lateral pile response. Computes and Geotechnics, under review (2004)

Poulos, H.G.: Behavior of laterally loaded piles II. Pile groups. J. Soil Mech. Found. Div (1971)

Randolph, M.F., Jewell, R.J., Poulos, H.G.: Evaluation of pile lateral load performance. In: Jewell and Andrews (eds.) Proceedings of Engineering for Calcareous Sediments, Balkema, Rotterdam, pp. 639–645 (1988)

Randolph, M.F., Houlsby, G.T.: The limiting pressure on a circular pile loaded laterally in cohesive soil. Geotechnique **34**(4), 613–623 (1984)

Zheng, W.: Horizontal static load test research for pile lateral friction and tip resistance of grouting technique. Chinese J. Rock Mech. Eng. (2015)

Ashour, M., Norris, G., Pilling, P.: Lateral loading of a pile in layered soil using the strain wedge model. J. Geotech. Geoenvironmental Eng. **124**(4), 303–315 (1998)

Ardalan, H., Ashour, M.: Application of the strain wedge model in soil-pile interaction analysis of pile-stabilized slopes. In: Geo-Congress 2013: Stability and Performance of Slopes and Embankments III, ASCE, pp. 1944–1955 (2013)

Modeling and Analysis of Soil-Pile Interaction for Dynamic Loading-A Review

Salman Ali Suhail[1(✉)], Fawad Ahmed Najam[2], and Adnan Nawaz[3]

[1] Geotechnical and Earth Resources Engineering, Asian Institute of Technology (AIT), Bangkok, Thailand
engrsalman07@gmail.com
[2] Structural Engineering, Asian Institute of Technology (AIT), Bangkok, Thailand
fawad.ahmed.najam@ait.asia
[3] Civil Engineering Department, COMSATS Institute of Information Technology, Wah Cantt, Pakistan
haveadnan@gmail.com

Abstract. A significant increase in the construction of high-rise structures have been observed in the last two decades, to accommodate the growing urbanization and increasing population needs. A considerable number of these high-rise structures is pile-supported and is located in seismically active areas. This urges the need for developing improved analytical techniques for the evaluation of complex dynamic response of pile-supported high-rise structures. Dynamic analysis of a soil-pile interaction system is one of the most complex problems in the field of geotechnical engineering. It involves the detailed considerations of various simultaneously occurring and coupled phenomena including the modeling of soil-pile interaction, pile group interaction factors and inertial interactions. A clear physical insight of soil-pile-structure interaction is of primary importance for accurate prediction of the true nonlinear behavior and response of the structures subjected to dynamic loading. This article is an attempt to review some appropriate tools to model the soil-pile interaction with more emphasis on the modeling of soil media. Various approaches, their applicability, practical considerations and limitations are also discussed. The presented information is intended to assist practicing geotechnical engineers in selection of appropriate modeling scheme for research and practical design purposes.

Keywords: Soil-pile interaction · Modeling of soil media · Dynamic soil response · Analysis of soil-pile systems

1 Introduction

The behavior of any structural system depends on the behavior of its key components. Generally, for civil engineering structures, the complete structural system comprises of 3D frames in superstructure, its foundation and the surrounding soil. In-depth understanding of the interaction among the aforementioned components is of utmost importance for an efficient, safe and economic design of overall system. Due to rapid increase in the urbanization, the construction trend has been shifted towards high-rise

T. Abdoun and S. Elfass (eds.), *Soil Dynamics and Soil-Structure Interaction for Resilient Infrastructure*, Sustainable Civil Infrastructures, DOI 10.1007/978-3-319-63543-9_12

buildings, widespread road networks, and long span bridges etc. In most of these structures, generally, piles are selected as an appropriate foundation type to resist, the static as well as the anticipated dynamic loading. Therefore, for the accurate predictions of design demands, a thorough understanding of soil-pile interaction is of vital importance. The soil-pile interaction comprises of two phenomena i.e., kinematic interaction and inertial interaction. The effect of ground motions on the foundation is referred to as the kinematic interactions, whereas the effect of superstructure loads on foundation is termed as inertial interactions (Fig. 1). Under dynamic loading, both interaction phenomena are needed to be considered in the analysis for an appropriate response prediction of soil-pile interaction (Peiris 2014).

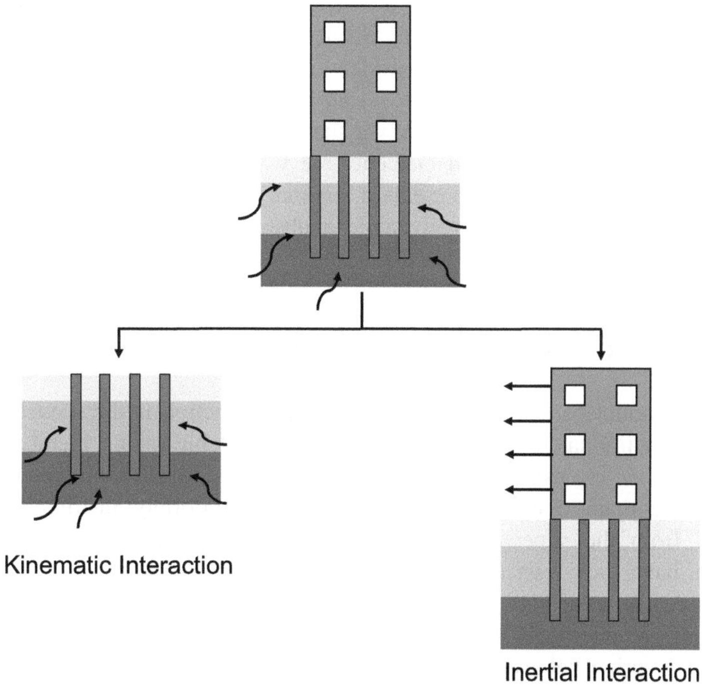

Fig. 1. Kinematic and inertial interaction

For the dynamic analysis and design of structures, a common practice is to model the structure in such a way that its movement at the base is restricted. However, in actual scenario, there is always some flexibility at the supports which decreases the stiffness of the system and enhances the natural period. Depending upon the type of soil/site and the characteristics of anticipated ground motions, the design response spectrum may show a significant variation in the spectral acceleration with slight change in natural period of the overall system. In this way, an increase in natural period can significantly alter the overall dynamic response of soil-pile interaction. Also, apart from affecting the shift in natural period, different soil media can exhibit different damping characteristics. These issues remained a subject of immense research during the last two decades (Jonathan et al. 1999; Stewart et al. 1999). In general, for a more realistic and physical insight, the combined soil-pile system should be modeled from the perspective of structural mechanics. Therefore, the selection of appropriate modeling and analysis techniques and methods is of great importance for accurate prediction response of soil-pile system.

The classical modeling approach for pile foundations is using the traditional beam theory (i.e., either Euler beam or Timoshenko beam). In Euler beam element, shear deformations are ignored and the mathematical formulation is based on the assumption of *plane sections remain plane and normal to the longitudinal axis*. The deformation at a particular section is rotation due to bending only. As transverse shear stress is not considered in Euler beam element, it provides good results for the normal stress and is proficient in apprehending the bending-controlled deformation fields. On the other hand, Timoshenko beam element considers the shear deformations, however, the analysis is still based on *plane sections remain plane* assumption but it is no longer normal to the longitudinal axis. Transverse shear stresses are taken into consideration and therefore, Timoshenko beam elements are widely used to model thick beams (Goerguelue 2009). This paper presents a summary of various modeling techniques available in the literature for the dynamic analysis of soil-pile system, with more emphasis on the modeling of soil media.

2 Classical Approaches for the Modeling of Soil Media

The most important part in modeling of soil-pile interaction is the modeling of soil media. Two traditional modeling approaches are available in the literature for this purpose, (a) modeling using the Winkler model (1867) and, (b) the continuum-based approach. The main objective of foundation is to uniformly distribute the loads transferred from the superstructure to the soil. Considering this viewpoint, both approaches have their own advantages and limitations. The basic principle and the merits and de-merits of both approaches are discussed below.

The basic Winkler (1867) model idealizes the soil as identical and linearly elastic springs. Each spring behaves independently and deforms only under the loaded area (Fig. 2). The elastic load-deformation relationship is given by the following form of Hook's law.

$$p = kw \tag{1}$$

Where p is the pressure, k is the coefficient of subgrade reaction and w is the deformation. The precise evaluation of subgrade modulus is important, since it is the only parameter that governs the behavior of soil when Winkler approach is used. Few experimental approaches used to determine the value of subgrade reaction are the plate load test, consolidation test for cohesive soils, tri-axial test and the California bearing ratio test (Terzaghi 1955). The major limitation of Winkler's hypothesis is that continuum material is modeled using independent springs and hence the interaction among soil particles under the applied loading is not considered. Therefore, the effect of external load becomes confined. Furthermore, springs used to model soil are linear elastic whereas under dynamic loading, soil behaves nonlinearly which is another serious limitation of Winkler model. To address these limitations, many researchers proposed refined/improved Winkler models which are illustrated in the later sections of this article.

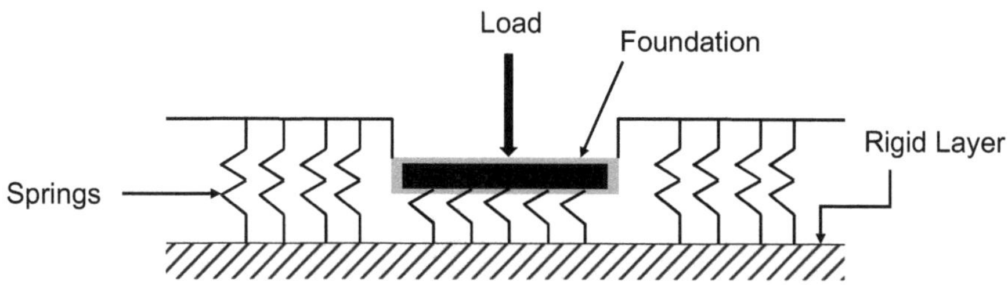

Fig. 2. Winkler foundation model

The second approach of modeling soil media is the continuum-based approach. Bossinesq (1885) was among the first to analyze the infinite, homogenous and elastic solid under normal force using classical theory of elasticity. The soil mass was assumed to be composed of small particles connected through inter-granular forces. The problems commonly dealt in soil mechanics involve very large boundary distances and loaded areas compared to the size of the individual soil grains. Hence, in effect, the body composed of discrete molecules gets transformed into a 'statistical macroscopic equivalent' amenable to the mathematical analysis. Therefore, continuum mechanics theory is much more appropriate to model soil media. In this approach, soil is modeled as semi-infinite and isotropic material in order to make the model less complex. An important advantage of this modeling approach is simplicity of input parameters since soil elastic modulus (E) and the Poisson's ratio (ν) are the only parameters required. However, continuum-based approach of soil modeling also has limitations in solving practical problems and it may lead to inaccurate evaluation of reactions at the edges of the foundations. The imprecise prediction of surface displacements away from the loaded region is also one of the limitations of this approach (Fletcher 1971). Various limitations in these discussed classical modeling techniques opened the doors for the researchers to develop improved models to represent soil media, as discussed in next section.

3 Improved Models for Foundations and Soil

In order to address the shortcomings of the classical modeling, various modified and improved foundation models were proposed for both the Winkler approach and the continuum-based approach. In general, two improvement techniques are employed by most of the improved models. In the first technique, structural elements are introduced to develop continuity in soil springs, whereas in the second approach, continuum-based models are simplified to develop more realistic soil representation.

3.1 Improved Versions of the Winkler Model

(a) *Filonenko-Borodich Foundation Model:* In this approach, an elastic membrane is introduced in the basic model to develop the connectivity in the Winkler springs. This elastic membrane is subjected to tension (T) and is attached to the top of the springs (Fig. 3). The governing behavior of this model is given by Eq. (2). The behavior of introduced elastic membrane accounts for the interaction among the springs, which was not considered in the classical modeling approach (Filonenko 1940).

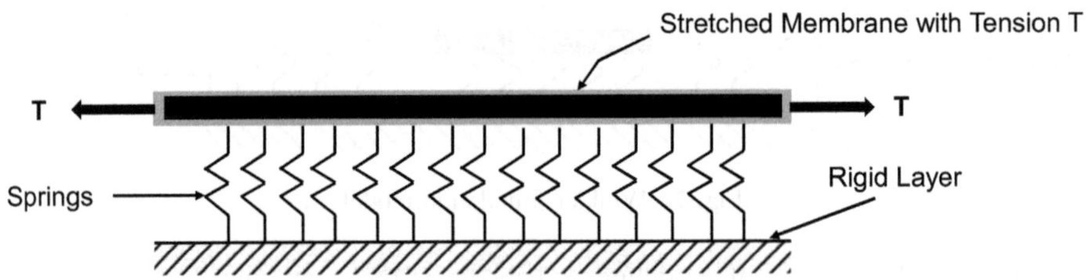

Fig. 3. Filonenko-borodich foundation model

For rectangular or circular foundations:

$$p = kw - T\nabla^2 w \tag{2}$$

For strip foundations:

$$p = kw - T\frac{d^2w}{dx^2} \tag{3}$$

Where, T is the tensile forces and ∇ is Laplace operator.

(b) *Hetenyi Foundation Model:* Hetenyi (1950) proposed a model in which, the interaction among the discretized springs is obtained by introducing an elastic beam/plate. However, this elastic beam/plate endures flexural deformations only (Fig. 4). The mathematical representation of Hetenyi foundation model is given in Eq. (4). This foundation model is considered a reasonable compromise between the Winkler foundation and the continuum-based approach, and is widely used for dynamic soil-pile interaction problems (Hetenyi 1950).

$$p = kw - D\nabla^4 w \tag{4}$$

Here D is the flexural rigidity of the plate and can be evaluated using Eq. (5).

$$D = \left(E_p h_p^3\right) / \left(12(1 - \mu_p)^2\right) \tag{5}$$

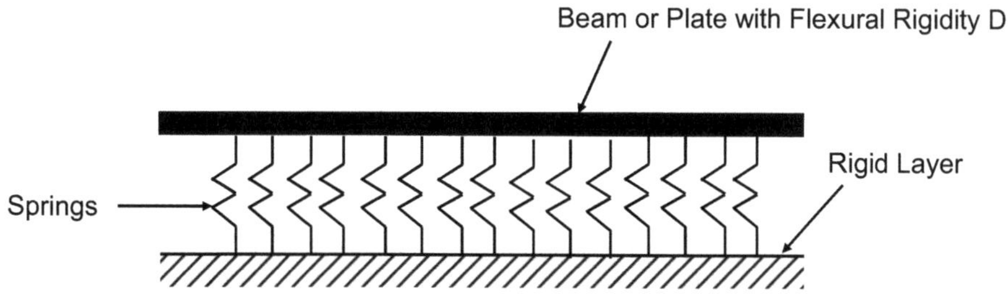

Fig. 4. Hetenyi foundation model

Where; E_p is the Young's modulus of the plate material, h_p is the plate thickness and μ_p is the Poisson's ratio of the plate material.

(c) *Pasternak Foundation Model:* Pasternak's (1954) foundation model can be considered as modified version of Hetenyi (1950) foundation model as it also incorporates the shear interaction mechanism among the spring elements by introducing beam/plate element that undergoes shear deformations only (Fig. 5). Vertical equilibrium is considered and pressure-deflection (p – w) relationship is defined by Eq. (6) (Pasternak 1954).

$$p = kw - G\nabla^4 w \tag{6}$$

Here G represents shear modulus of newly introduced shear layer.

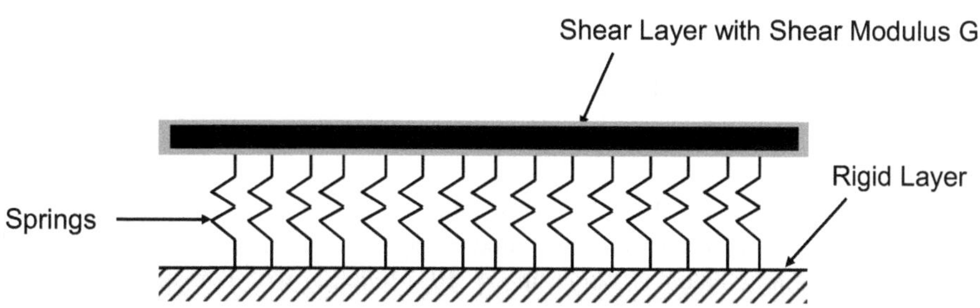

Fig. 5. Pasternak foundation model

(d) *Generalized Foundation Model:* An additional assumption is made to Winkler hypothesis i.e., at each contact point, the moment is directly proportional to the rotation angle. This additional assumption leads to the development of the generalized foundation model. The mathematical representation is given by Eqs. (7) and (8). The assumption made here is rather arbitrary, however the physical implications of this model are established in literature (Kerr 1964)

$$p = kw \tag{7}$$

$$m_n = k_1 \frac{dw}{dn} \tag{8}$$

Where; m_n is the moment in n direction, k_1 is the proportionality factor.

(e) *Kerr Foundation Model:* Shear surface/layer is familiarized in Winkler foundation model. In this mode, the coefficient of springs above and below this surface/layer are anticipated to behave differently (Fig. 6). Mathematical representation of Kerr foundation model is shown in Eq. (9) (Kerr 1965).

$$\left(1 + \frac{k_2}{k_1}\right)p = \frac{G}{k_1}\nabla^2 p + k_2 w - G\nabla^2 w \tag{9}$$

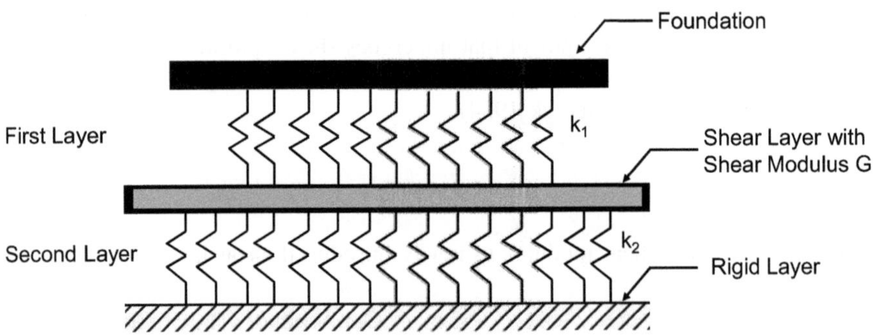

Fig. 6. Kerr foundation model

(f) *Beam Column Analogy Model:* A new subgrade model is introduced in the classical *"beam on elastic foundation"* model (Fig. 7) and the prevailing differential equations are articulated as shown in Eq. 10. Modeling with this approach shows significant improvement over Winkler hypothesis, however, it may also lead to inaccurate results due to introduction of a fabricated shear force in the model.

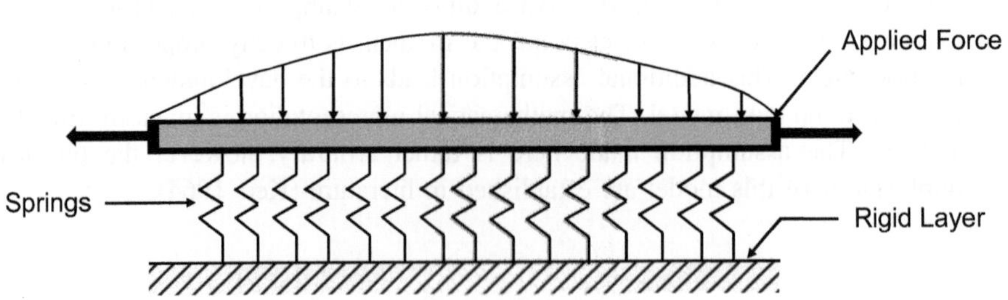

Fig. 7. Beam column analogy

$$E_b I_b \frac{d^4 w(x)}{dx^4} - C_{p2} \frac{d^2 w(x)}{dx^2} + C_{p1} w(x) = q(x) \tag{10}$$

Where $E_b I_b$ is the flexural stiffness of the beam, $w(x)$ is the settlement of beam, $q(x)$ is the load applied, C_{p1} and C_{p2} are constants (Horvath 1993).

(g) *New Continuous Winkler Model:* An idea was proposed to incorporate interactions among springs by intermeshing of springs instead of introducing new structural element in the model. The meshing is carried in such a way that the interconnection among the springs is achieved (Fig. 8). This powerful idea leads to the development of new continuous Winkler model. Physically, interaction among springs without additional structural element is achieved by adding another spring having specific axial stiffness which is not in contact with the foundation. More details and the case studies for this model are available in the literature (Kurian 2001).

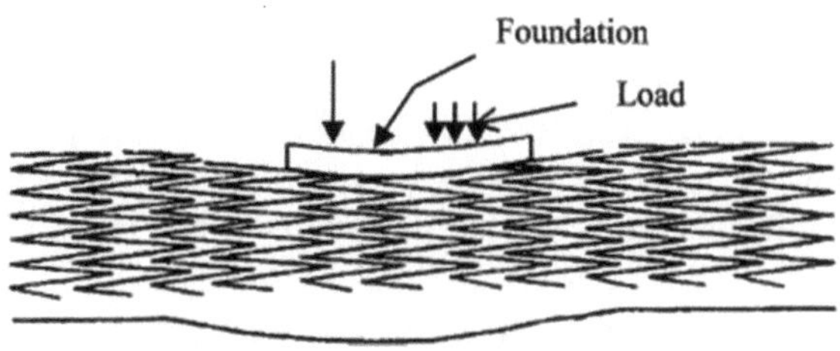

Fig. 8. New continuous winkler model

3.2 Improved Versions of the Continuum Model

(a) Vlasov Foundation: The variational principle had been adopted in continuum approach to develop a foundation model that represents soil-structure interaction in a more realistic way. This model limits the deformations of an elastic layer by introducing a certain function in the formulation, as illustrated below:
   (i) Vertical displacement is computed as function of $h(z)$ which governs the variation of deformation in vertical directions. Vertical displacement is computed using Eq. (11).

$$w(x.z) = w(x).h(z) \tag{11}$$

   (ii) Horizontal displacement is assumed to be zero.

   The pressure-deformation equation of Vlasov foundation model is developed using the principle of virtual work and is given by Eq. (12) (Vlasov 1949; Vlasov et al. 1960).

$$p = kw - 2t\frac{d^2w}{dx^2} \tag{12}$$

Where;

$$k = \frac{E_o}{(1 - v_o)^2}\int_0^H \left(\frac{dh}{dz}\right)^2 dz \tag{13}$$

$$t = \frac{E_o}{4(1 - v_o)}\int_0^H h^2 dz \tag{14}$$

$$E_o = \frac{E}{(1 - v^2)} \tag{15}$$

$$v_o = v(1 - v) \tag{16}$$

(b) *Reissner Model:* This model is based on two assumptions, (a) the plane stresses in the foundation are very small, hence can be ignored, and (b) the horizontal stresses are zero (same as in Vlasov foundation model). The pressure-deformation relationship at the interface between foundation and subgrade is determined as disturbance of foundation below the slab and is shown in Eq. (17).

$$C_1 w - C_2 \nabla^2 w = p - \frac{C_2}{4C_1}\nabla^2 p \tag{17}$$

Where C_1 and C_2 are the constants. Several other improved continuum-based models are also proposed and are available in the literature (Harr et al. 1969; Rao et al. 1971; Nogami et al. 1987; Horvath 1989; Vallabhan et al. 1991).

With the development of advanced computational technologies, the determination of relative accuracy of various models to idealize the soil media in soil-pile interaction, is now an active research area. It is worth mentioning here that the overall behavior of soil-pile system may heavily depend on the modeling scheme and the uncertainties involved in determining the governing parameters used in model. The foundation models discussed in this article require the determination of various parameters e.g. modulus of subgrade reaction, soil damping coefficients, index and strength properties etc. These parameters can be determined using field or laboratory testing of the soil. The complexity of any model increases with the number of required input parameters. The Winkler model (for representing soil) offers the most practical convenience as it only requires the modulus of subgrade reaction, which can be easily determined using the plate load test. The value obtained can be conveniently modified for actual foundation size, shape etc. (Terzaghi 1955). Other required parameters can be obtained using the rigid stump test (Siva et al. 1969; Pranesh 1970; Siva et al. 1971; Siva et al. 1973; Vallabhan et al. 1991). The major shortcoming of Winkler approach is the lack of interaction among the soil springs, as it models the soil media using independent springs. In the case of continuum-based modeling approach, the young's modulus of soil is the necessary parameter. Since an accurate determination of the young's modulus of soil is relatively complex, the Winkler modeling approach gained more appreciation among the practicing geotechnical engineers (Bowles 1997).

4 The Complex Soil Behavior and Their Modeling

The response of soil to dynamic loading is complex enough due to a number of reasons. As discussed earlier, soil is composed of granular particles that are joined together by intergranular forces. The stress-strain characteristics of soil govern the overall dynamic soil-pile interactions. Under a general loading, the soil mass tends to attain explicit structural configurations so that the potential energy becomes minimum. Under certain level of loading, the soil stress-strain behavior remains elastic, after then it enters into the plastic range, with an unproportioned increase in deformation with increasing loads. This soil deformation is followed by viscoplastic deformation which is dominant in fine grained soil. This initial deformation in the soil is mostly occurred due to the expulsion of pore water under the loading while simultaneously transferring the pore pressure to the solid soil mass. As a result of this behavior, the rate of strain decreases with time. This whole process of expulsion of water from the soil under the action of loading and squeezing of soil mass is termed as the primary consolidation. After attaining primary consolidation, the soil still resists load for certain interval of time and secondary compression of soil occurs following the logarithmic function of time (Bowles 1997). It must be noted that the behavior of unsaturated soil having sufficient elastic range will be different, as the strain will no longer be the function of time. On the other hand, for a fully saturated soil sample, the strain will be a function of time under dynamic loading. This hardening behavior of fine grained saturated soil must be incorporated in the model for the accurate prediction of response of soil-pile structure system, since this time-dependent property causes a reasonable increase in the strength characteristics of the soil (cohesion and angle of internal friction). Furthermore, under dynamic loading, the stress-strain behavior of soil no longer remains linear, as the soil exhibits inelastic behavior. Hence, the nonlinearity of soil must also be considered in the modeling and analysis of the dynamic soil-pile interaction problems.

4.1 Elasto-Plastic Idealization

The idealized elasto-plastic behavior is one of the most common ways to incorporate soil nonlinearity in the soil-pile interaction model. There are numerous ways to model elasto-plastic behavior of soil. One of the ways is the introduction of elasto-plastic element in the model. Hookean spring element is the most appropriate elasto-plastic element. The Coulomb element (Zeevart 1972), which efficiently represents the plastic deformation of soil, is used in conjunction with the Hookean spring element to model

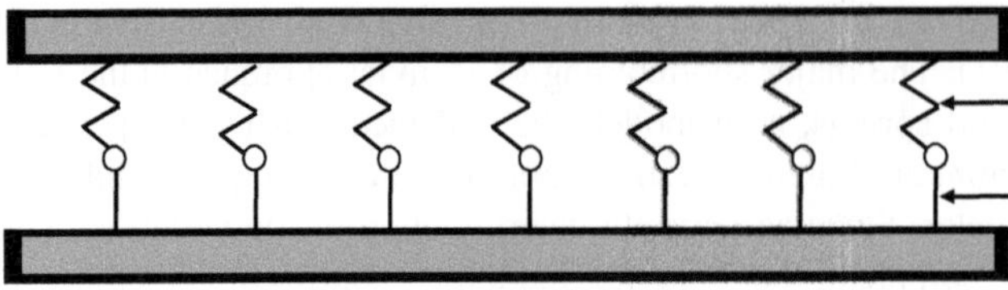

Fig. 9. St. Venant's unit

elasto-plastic behavior of soil. This conjunction forms the schematic system referred to as the St. Venant's unit (Fig. 9). It is worth mentioning that St. Venant's unit should be used in large number and connected in parallel for smooth transition of soil from elastic to plastic state. In case if only one element is used, the transition is abrupt and the accuracy of results can be compromised (Dutta et al. 2002). The mathematical expression for strain moduli for elastic and plastic range is shown in Eq. (18):

$$\varepsilon_{ep} = M_e\sigma + M_p log\frac{\sigma_u}{\sigma_u - \sigma} \tag{18}$$

Where; M_p and M_e are the plastic and elastic strain moduli, respectively, and σ_u is the ultimate load. This model appears to be useful, however, practical problems can occur with regard to the selection of input parameters and arrangement of springs. Numerous yield criterion such as Von Mises, Mohr-Coulomb, Drucker-Prager and Tresca yield criterion (Harr 1966; Das 1983) had been available in the literature, which can be adopted for modeling elasto-plastic idealization of soil. The post-yielding behavior can be incorporated in the model using either deformation theory, flow theory or incremental theory (Desai et al. 1987). In deformation theory, plastic strains are uniquely defined by the state of stress, whereas in incremental theory, plastic strains are dependent on multiple factors including the state of stress, increment in stress and strain etc. Generally, the incremental theory is applied in modeling the post-yielding behavior. The constitutive modeling of soil can be developed using one of the following methods; iterative method, initial strain method, initial stress method and incremental method. The detailed formulations are available in the literature (Desai et al. 1987; Ziendkiewicz et al. 1989). The key advantage of these formulations includes relatively simple coding of yield function and flow rule, as only simple soil parameters (which can be easily obtained) are required to implement the aforementioned formulations in computer programs. The 3D elasto-plastic model for the unsaturated soil is also proposed recently (Sun et al. 2000).

4.2 Viscoelastic Idealization

The permeability of soil is a vital parameter governing the deformation characteristics of saturated fine grained soils. As mentioned earlier, the water in pores is expelled under the loading, resulting in soil deformation and pore water pressure on soil mass. The pore water pressure results in a time-dependent, progressive growth in effective stress of soil, which ultimately leads to the time-dependent settlement of foundation.

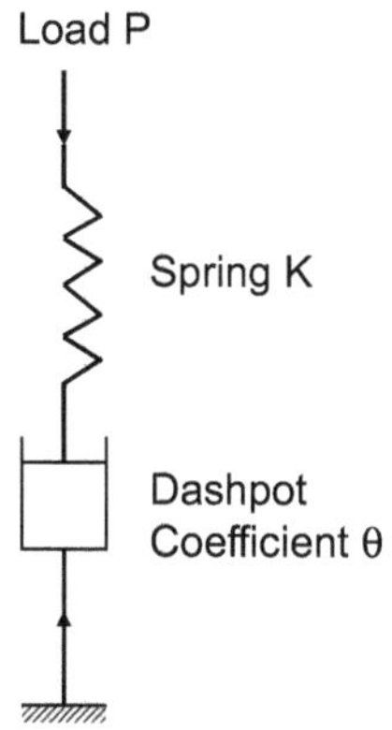

Fig. 10. Maxwell element

Various models attempting to idealize the rheological characteristics of clayey soil are well documented in the literature. These include mechanical models, theory of plastic flow and the molecular theory of flow (Vyalov 1978). The elastic element in conjunction with viscous and plastic elements is a common way to model rheological soil properties. Springs and dashpots are used in combinations for this purpose. These springs and dashpots can be connected either in series i.e. Maxwell model (Fig. 10) or in parallel i.e., Kelvin model (Fig. 11). Various conceptual details and explanation of these models are available in the literature (Vyalov 1978).

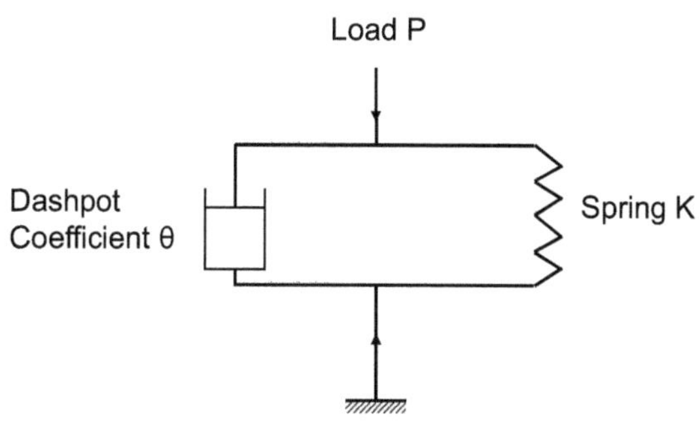

Fig. 11. K element

5 A Review of Various Analysis Approaches for Soil-Pile Interaction System

The complete analysis of soil-pile interaction system involves the modeling and analysis of piles and pile cap, the selection of boundary conditions, the modeling of soil media, soil-pile interaction, non-linear soil response and various other important components and phenomenon. Various analytical methods, based on idealization of soil behavior, were developed over the past few decades to predict the response of soil-pile system subjected to dynamic loading. The analytical methods can be categorized into two types i.e. the "beam on elastic foundation" approach and the "continuum-based" approach. Their brief introduction and underlying concepts are reviewed as follows.

5.1 *Beam on Foundation* Approach

In this analytical method, the soil is modeled as a series of discretized springs and their stiffness (k) defines the soil strength (Fig. 12), while the pile is modeled as a flexible beam. It is based on the basic idea proposed by Winkler i.e., the subgrade resistance in vertical direction against any external force is directly proportional to the ground deformation (Basu et al. 2008). In 1946, Hetenyi presented static solution of laterally loaded piles by adopting Winkler model. The closed-form solution was obtained by considering elastic spring resistance as a function of pile lateral displacement (Hetenyi 1946). This leads to the formulation of standard framework of analysis of piles based on Winkler model. Following this study, various analytical solutions based on Winkler method have been proposed and reported in the literature (Glesser 1953; McClelland 1956; Matlock et al. 1961; Wood 1979). The deflection of the beam can be determined by the fourth-order governing linear differential equation as shown in Eq. (19). Hetenyi (1946) presented one of the earliest studies based on the closed-form solution of this equation by considering elastic spring resistance as a function of pile lateral displacement.

$$E_p I_p \frac{d^4y}{dx^4} + Q\frac{d^2y}{dx^2} + ky = p(x) \tag{19}$$

Where; E_p and I_p are the elastic modulus and moment of inertia of pile foundation, respectively, Q is the axial load, x is the vertical depth and y is the lateral deflection of the pile at a depth x. It is worth mentioning here, that in the case of foundation springs, the spring deflection multiplied by the spring constant results into resistive force of the ground, unlike the conventional spring which provides the spring force. Using this approach, various important characteristics such as the deflection of beam, bending moment and shear force can be evaluated. The basic parameters required to calculate

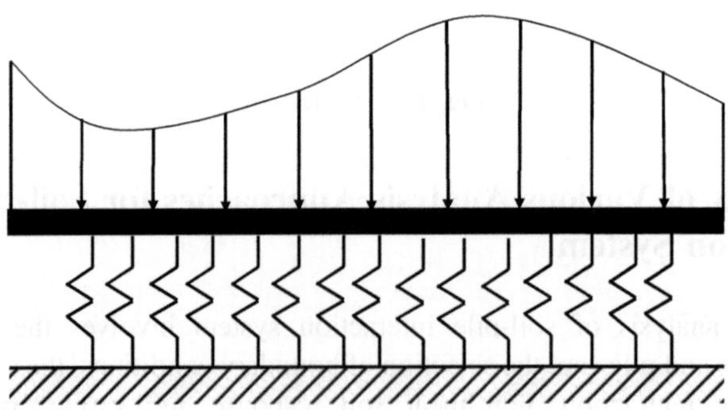

Fig. 12. Beam on elastic foundation

the aforementioned characters are elastic modulus, beam dimensions, loading conditions and the spring constant. The spring constant can be obtained conveniently using empirical methods. Due to its simplicity, the *beam on foundation* approach has been widely used for the dynamic analysis of soil-pile interaction. However, the prediction of dynamic behavior of soil-pile interaction is relatively complex due to soil nonlinearity (Matlock et al. 1960). Also, several important factors are ignored in this approach. These include soil-pile gapping, soil slippage, damping and analysis over the wide range of frequencies, which are not possible to consider using elastic springs on one side of the pile foundation.

Various refinements in the *beam on foundation* approach were proposed by the introduction of dashpot in addition to springs (Fig. 13). Due to the addition of the dashpot, the dynamic response of the soil-pile interaction can be obtained over a wide range of frequencies (Nogami et al. 1989). Furthermore, to account for the nonlinearity of soil media, soil is divided into two regions i.e., near field element and far field element. Near field element accounts for the soil nonlinearity, whereas the far field element governs the elastic behavior (Fig. 14) (Nogami et al. 1992). To account for the soil-pile gapping and soil slippage phenomena, the near and far field element concept was extended and used at both sides of pile foundation. The near field element comprises of nonlinear springs and stiffness is calculated assuming that the plane stress conditions govern. The near field element is modeled by a homogenous isotropic viscoelastic material. The second part is far field model, which accounts for the wave propagation away from the pile.

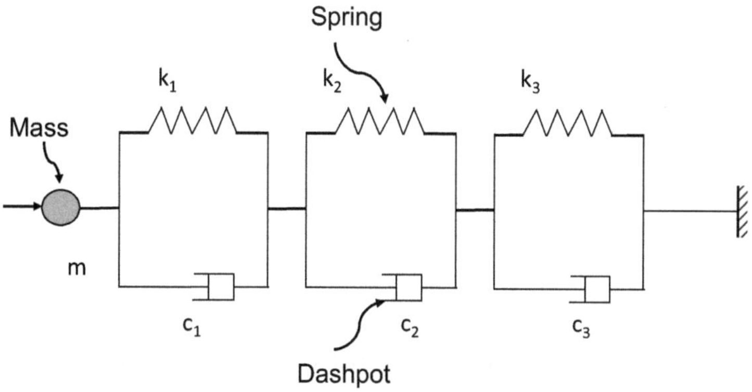

Fig. 13. (a) Winkler model (b) Near and far field element & interface model

A modified version of *beam on foundation* approach commonly known as the "$p - y$" method was also proposed, in which the pile is modeled using Euler beam element and soil is treated as a series of springs. Here p defines the soil resistance whereas y is the pile deflection against loading (Ashour et al. 2000; Prezzi et al. 2009). The $p - y$ method consents the pile response to be evaluated while considering the soil nonlinearity (Anderson et al. 1999). The *beam on foundation* approach was extended by incorporating the inelastic $p - y$ curve in the modeling and hence the elastic

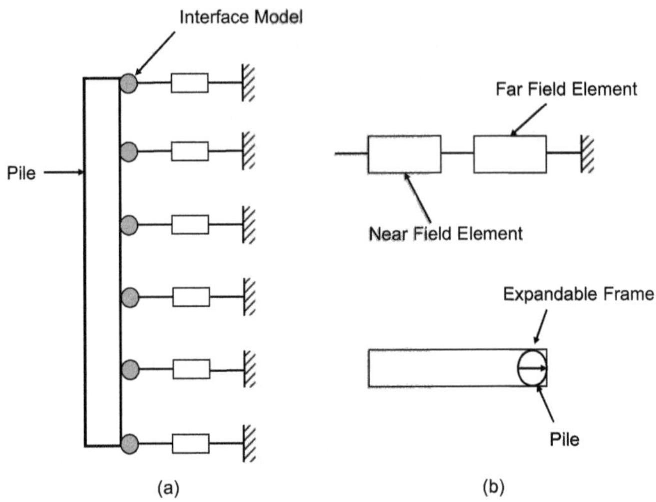

Fig. 14. Spring and dashpot

foundation is replaced by nonlinear foundation. In beam on elastic foundation methods, the spring constant was used as an input for the analysis, however in this approach, the complete load-deflection $(p-y)$ curve is required as an input parameter. The prediction of complete cyclic behaviors can also be accurately obtained using the detailed hysteretic models. A generalized hysteretic model has four major components i.e., backbone curve, unloading curve, reloading curve and cyclic degradation rule. The basic soil load-deflection curve is generally used as backbone curve. It is important to note that the accuracy of results using this model primarily depends on the adequacy of the basic $p-y$ relationship. Special care must be taken in order to obtain the accurate load-deflection curve for the given site (Allotey et al. 2008a, 2008b). The strain wedge method was introduced to generate $p-y$ curves. With the introduction of this method, various important parameters such as the 3D soil–pile interaction, soil continuity, cyclic strength, degradation of stiffness, soil slippage and gapping phenomenon, are taken into consideration using the beam on nonlinear foundation approach. The traditional ways of determining the $p-y$ relationships may result in some shortcomings such as the lack of consideration of soil continuity, material stiffness, cross-section shape and head conditions (Heidari et al. 2014). The $p-y$ method is considered as a useful tool as compared to the Winkler method, since it indirectly considers the soil continuity in the analysis. However, there is no explicit method in $p-y$ analysis approach to incorporate soil as continuum material. Care must be taken while extrapolating the $p-y$ curves especially in cases where no experimental tests are performed, as a small amount of error may lead to significantly inaccurate results (Chen et al. 2014)

5.2 Continuum-based Methods

The second approach is *continuum-based approach*, in which the soil media is treated as a three-dimensional continuum material unlike *beam on foundation* approach which treats soil as one-dimensional discretized springs. This three-dimensional treatment of soil media is more tempting as it helps in considering the 3D aspects of interaction among the soil and piles. One of the earliest continuum-based analysis methods is the finite difference method (FDM). It was developed in late 1760 s but gained more importance in mid 1920 s when stability conditions were presented for hyperbolic partial differential equations. The FDM is based on discretization technique to solve different sets of governing differential equations. With recent advancements in computer applications and information technology, the FDM is also employed to study the soil-pile interactions, modeling of reinforced earth, analyzing the inherent capacities of stone columns, soil liquefaction and tunneling etc. For dynamic analysis of soil-pile interaction, various FDM computer codes were written (such as FLAC) in which 2D and 3D dynamic analysis of soil-pile interaction can be carried out. Various formulations of computer codes and their applications are well documented in the literature (Chau 2013). Another commonly known continuum-based method for the dynamic analysis is the boundary element method (BEM). In the BEM, the soil-pile interaction is undertaken by apportioning the soil and pile into number of slices. The interaction among different soil and pile slices and the analysis can be done by adopting Mindlin solutions (Mindlin 1936). The well-known finite element analysis (FEA) is another continuum-based analysis approach and is considered as the most versatile and well-established method for the dynamic analysis of soil-pile interaction. Its automation in various programming environments and efficient numerical solvers have currently made it the most widely-used analysis method. It is a discretization approach in which the soil and pile are divided into finite number of elements. Various basic principles and applicability of the FEM for soil-pile interaction system, considering soil nonlinearities and geometrical changes, are well documented in the literature. (Desai et al. 1982, 1987; Zienkiewicz et al. 2005). Owing to its wider applicability and ability to handle complex physical interactions, the FEM approach will be discussed in more detail in next section.

6 The Finite Element Modeling and Analysis For Soil-Pile Interaction System

With recent advancements in computational technology and powerful computers, the use of the finite element method (FEM) to model and analyze the complex physical phenomena is rapidly increasing. In the case of soil media, the mechanisms such as soil nonlinearity, non-uniformity, discontinuities, complex geometrical changes etc., can be modeled and analyzed using the FEM. Similarly, the viscoelastic behavior of soil media can also be expediently modeled and analyzed using the finite element method (Thomas et al. 1989). The basic principle of the FEM is based on variational approach, where the whole system is discretized into finite elements connected at different nodes. The stiffness matrix of the whole system is determined using stiffness matrices of each

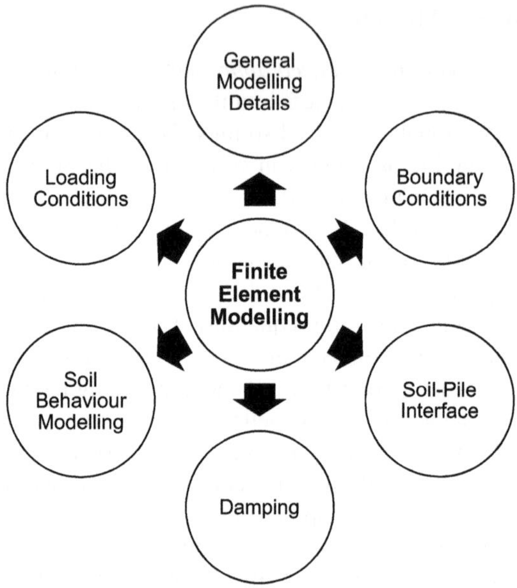

Fig. 15. Components of a finite element model

finite element. The unknown displacements are evaluated using global loading conditions and boundary conditions. For the purpose of dynamic analysis of a soil-pile interacting system, the components of a typical finite element model are shown in Fig. 15.

The general finite element modeling starts with the selection of appropriate finite elements to represent the soil and pile. For 3D soil-pile interaction models, generally 8-noded brick elements are used to model the soil media. The boundary conditions are used in dynamic analysis to reduce the analysis domain. Unlike static analysis, the fixed boundary conditions cannot be applied at some distance from the interested region, as it can reflect outward propagating waves back to the model and hence can affect the accuracy of results. Therefore, the selection of appropriate boundary conditions is quite important. Various options for the boundary conditions are available such as the transmitting boundary conditions, absorbing boundary conditions, continued fraction absorbing conditions etc. The detailed discussion on the boundary conditions is beyond the scope of this review. The adequacy of interface modeling between the soil and pile can also significantly affect the dynamic response. It can be done in two ways i.e., the perfectly bonded interface or frictional interface. The two important parameters which must be considered while modeling the interface between soil and pile are the soil slippage and soil gapping. The frictional modeling of interface is a good option since it directly incorporates the effects of soil slippage and gapping in the analysis. Various studies are available in the literature to develop interface elements which can account for discontinuity at the interface (Goodman et al. 1968; Zienkiewicz et al. 1970; Ghaboussi et al. 1973; Pande et al. 1979; Beer 1985).

With the propagation of waves in the soil media, their energy dissipates due to the inherent damping offered by soil. For dynamic analysis using the FEA, two types of damping are considered i.e., the material damping and geometrical/radiation damping. The magnitude of damping should be appropriately selected based on the empirical investigations and existing data. Similarly, the selection of appropriate constitutive soil model is also a pre-requisite for the accurate dynamic analysis of soil-pile interaction problems. Two types of material models (linear/nonlinear elastic and elasto-plastic) are generally used in the FEA. It is imperative to apply the initial loading conditions, e.g. the dead loads and other static loads of similar nature, prior to the dynamic loading on the model. These initial loading conditions should represent the actual loading that foundation would resist in the absence of any dynamic load, and thus, should provide a right starting point for the lateral load analysis.

The earliest applications of the finite element method involved the determination of behavior of elastic pile considering the 2D soil models. Although the response of piles was evaluated considering soil nonlinearity and soil-pile slippage, the early studies were limited to kinematic interaction (Trochanis et al. 1991). Wu et al. (1997a; 1997b) performed the dynamic analysis of soil-pile interaction by considering both the kinematic and inertial interaction phenomenon. A quasi 3D finite element model was proposed which evaluates the time-dependent stiffness and damping factors in addition to the dynamic response of soil-pile system (Wu et al. 1997a, 1997b). The application of the FEM for dynamic analysis was extended by introducing the plasticity-based hierarchical single surface (HiSS) to model the soil media. Various important characteristics such as the deformation modes of bonding, soil slippage, pile-soil separation and rebounding of piles can be efficiently evaluated by adopting the combination of the FEM and the HiSS (Cai et al. 2000). The FEM was also employed to predict the dynamic response of socketed and floating piles as well as micro piles (Naggar et al. 2000; Sadek et al. 2003). The 3D finite element analysis was also used to investigate the effects of plasticity of soil (Maheshwari et al. 2004). The major disadvantage of performing dynamic analysis using the FEM is high computation cost. This shortcoming makes it less common in practical design applications. However, the reduced-dimension finite element formulations are also proposed recently to carry out the nonlinear analysis of vibrating discs in 2D unbounded domain, considering the soil nonlinearity. More recently, a simplified finite element formulation is presented as a fair compromise between the realistic representation of soil-pile interaction and the accuracy in numerical results (Thammarak et al. 2011).

7 Concluding Remarks

The review suggests that the *beam on foundation* approach remained the most widely used analysis approach for soil-pile interaction problems in last few decades. However, one of the primary shortcomings associated with this approach is that it does not considers the soil media as 3D continuum, rather it treats it as the one-dimensional discretized springs. Hence, the radial and 3D interaction components are not considered. Many improvements had been carried out to the traditional *beam on foundation* approach to evaluate the dynamic response of soil-pile interaction system, however, such improvements are not commonly used in practical design due to their complexity. The *beam on foundation* approach is also criticized for not accounting the interaction among the discretized springs, since the displacements at one point does not affect any other point in the analysis domain. On the other hand, continuum-based approach addresses most of the short comings associated with *beam on foundation* approach. This approach treats the soil as a continuum material and also considers the 3D interaction among the soil-pile-structure. The most common continuum-based method (i.e., the finite element method, FEM) is discussed in detail. The FEM have recently gained more popularity due to its generality and diverse options for modeling the complex conditions. The complex soil behaviors such as the material nonlinearity, non-homogenous material conditions, geometrical changes etc., can be modeled with reasonable accuracy. However, the use of FEM demands significant amount of computational cost to generate results with high precision. Based on the review of modeling and analysis techniques for dynamic analysis of soil-pile interaction, following are some observations.

(a) To anticipate the dynamic response of soil-pile interaction, the combined effect of kinematic and inertial interaction must be considered in the modeling and analysis.
(b) Although Winkler hypothesis have obvious shortcomings and limitations, it can still yield reasonably accurate results. For the practical point of view, Winkler idealization can be used in combination of modeling structures with fixed base.
(c) In the case of clayey soil, the consolidation behavior is of utmost importance. Under such conditions, the nonlinear stress-strain relationship must be considered for the dynamic analysis. The clayey soil has low permeability and time-dependent behavior under loading, which can be more accurately modeled using the viscoelastic idealization.
(d) The complete soil-pile interaction system can be modeled efficiently using the discretization techniques. The finite element modeling can provide reasonably accurate response predictions and is also quite useful to incorporate the nonlinearity, non-homogeneity and anisotropy of soil.
(e) The *beam on foundation* approach have obvious limitations for the dynamic analysis, but are still used widely for the design of soil-pile interaction systems. The main reason of its widespread usage is its simplicity and less computational cost. However, the accuracy of results can be lower compared to the detailed finite element analysis, which can now be regarded as the state-of-the-art practice.

This article is an attempt to briefly summarize the widely-used modeling and analysis methods for the dynamic analysis of soil-pile systems. Further details of any particular modeling technique or analysis method can be found in the bibliography list provided as the references.

References

Allotey, N., El Naggar, M.H.: A numerical study into lateral cyclic nonlinear soil–pile response. Can. Geotech. J. **1281**, 1268–1281 (2008a). doi:10.1139/T08-050

Allotey, N., El Naggar, M.H.: Generalized dynamic Winkler model for nonlinear soil–structure interaction analysis. Can. Geotech. J. **573**, 560–573 (2008b). doi:10.1139/T07-106

Ashour, M., Norris, G.: Modelling lateral soil-pile response based on soil-pile interaction. J. Geotech. Geo-environ. Eng. **126**(5), 420–428 (2000)

Basu, D., Salgado, R., Prezzi, M.: Analysis of laterally loaded piles in multilayered soil deposits. Joint Transportation Research Program, 151 (2008). doi:10.5703/1288284313454

Bowles, J.E.: (1997). Foundation Analysis and Design, 5th (edn.) Engineering Geology, vol. 20 (1997). doi:10.1016/0013-7952(84)90010-3

Cai, Y.X., Gould, P.L., Desai, C.S.: Nonlinear analysis of 3D seismic interaction of soil–pile–structure systems and application. Eng. Struct. **22**(2), 191–199 (2000). doi:10.1016/S0141-0296(98)00108-4

Chen, W.-F., Duan, L.: Bridge Engineering Handbook, 2nd edn. CRC Press, Boca Raton (2014). doi:10.1201/b15616

Cox, W.R., Reese, L.C., Grubbs, B.R.: Field testing of laterally loaded piles in sand. In: Offshore Technology Conference, pp. 459–464 (1974). doi:10.4043/2079-MS

Dutta, S.C., Roy, R.: A critical review on idealization and modeling for interaction among soil-foundation-structure system. Comput. Struct. **80**(20–21), 1579–1594 (2002). doi:10.1016/S0045-7949(02)00115-3

El Naggar, M.H., Novak, M.: Nonlinear analysis for dynamic lateral pile response. Soil Dyn. Earthq. Eng. **15**(4), 233–244 (1996). doi:10.1016/0267-7261(95)00049-6

Goerguelue, U.: Beam Theories: The difference between Euler-Bernoulli and Timoshenko. Lecture Handouts (2009). Accessed

Heidari, M., El, H., Jahanandish, M., Ghahramani, A.: Generalized cyclic p – y curve modeling for analysis of laterally loaded piles. Soil Dyn. Earthq. Eng. **63**, 138–149 (2014). doi:10.1016/j.soildyn.2014.04.001

Hetenyi, M.: A general solution for the bending of beams on an elastic foundation of arbitrary continuity. J. Appl. Phys. **21**(1), 55–58 (1950). doi:10.1063/1.1699420

Horvath, J.S.: Beam-column-analogy model for soil-structure interaction analysis. J. Geotech. Eng. **119**(2), 358–364 (1993)

Jonathan, P., Angeles, L., Raymond, B., Greggory, L., Uildings, B., Stewart, B.J.P., Fenves, G.L.: Seismic soil-structure interaction in buildings II: empirical findings. J. Geotech. Geo-environ. Eng. **125**, 38–48 (1999)

Kerr, A.D.: A study of a new foundation model. Acta Mech. **1**(1), 135–147 (1965)

Kerr, A.D.: Elastic and viscoelastic foundation models. J. Appl. Mech. **31**, 491 (1964). doi:10.1115/1.3629667

Maheshwari, B.K., Truman, K.Z., El Naggar, M.H., Gould, P.L.: Three-dimensional finite element nonlinear dynamic analysis of pile groups for lateral transient and seismic excitations. Can. Geotech. J. **41**(1), 118–133 (2004). doi:10.1139/t03-073

Mcgann, C.R., Arduino, P., Mackenzie-helnwein, P.: Simplified procedure to account for a weaker soil layer in lateral load analysis of single piles. J. Geotech. Geoenvironmental Eng. **138**, 1129–1137 (2012). doi:10.1061/(ASCE)GT.1943-5606.0000684

Mechanik, M.U.N.D.: Zeitschrift for Angewandte, 1 (1925)

Mindlin, R.D.: Force at a point in the interior of a semi-infinite solid. J. Appl. Phys. **7**(5), 195 (1936)

Naggar, M.H.E., Bentley, K.J.: Dynamic analysis for laterally loaded piles and dynamic p–y curves. Can. Geotech. J. **37**(6), 1166–1183 (2000)

Nogami, T., Konagai, K.: Time domain flexural response of dynamically loaded single pile. J. Eng. Mech. **114**(9), 1512–1525 (1989)

Nogami, T., Lam, Y.C.: Two-parameter layer model for analysis of slab on elastic foundation. J. Eng. Mech. **113**(9), 1279–1291 (1987)

Nogami, T., Otani, J., Konagai, K., Chen, H.: Nonlinear soil-pile interaction model for dynamic lateral motion. J. Geotech. Eng. **118**(1), 89–106 (1992)

Pande, G.N., Sharma, K.G.: On joint/interface elements and associated problems of numerical ill-conditioning. Int. J. Numer. Anal. Methods Geomech. **3**(3), 293–300 (1979)

Pasternak, P.L.: On a new method of analysis of an elastic foundation by means of two foundation constants. Gas. Izd. Lit. po Strait I Arkh, Moscow, Russia (in Russian) (1954)

Peiris, T.P.: Soil-Pile Interaction of Pile embedded in Deep Layered Marine Sediment under Seismic Excitation. Ph.D. Thesis, Queensland University of Technology (2014)

Prezzi, M., Salgado, R., Basu, D.: A continuum-based model for analysis of laterally loaded piles in layered soils. Géotechnique **59**(2), 127–140 (2009). doi:10.1680/geot.2007.00011

Sadek, M., Shahrour, I.: Influence of piles inclination on the seismic behavior of groups of flexible piles. Transp. Res. Rec. 1–14 (2003)

Stewart, J.P., Fenves, G.L., Seed, R.B.: Seismic soil-structure interaction in buildings I: analytical methods. J. Geotech. Geoenvironmental Eng. **125**, 26–37 (1999)

Terzaghi, K.: Evalution of conefficients of subgrade reaction. Geotechnique **5**(4), 297–326 (1955). doi:10.1680/geot.1955.5.4.297

Thammarak, P., Tassoulas, J.L.: Reduced-dimension nonlinear finite-element analysis of a vibrating disk in an unbounded domain. Earthq. Eng. Struct. Dynam. **40**(January), 1353–1366 (2011). doi:10.1002/eqe

Thomas, H.R., Bendani, K.: Primary/secondary compression solution algorithm. J. Comput. Civ. Eng. **114**(5), 578–593 (1989)

Trochanis, B.A.M., Member, A., Bielak, J., Christiano, P.: Three-dimensional nonlinear study of piles. J. Geotech. Eng. **117**(3), 429–447 (1991)

Vallabhan, C.V.G., Das, Y.C.: Modified vlasov model for beam on elastic foundations. J. Geotech. Eng. **117**(6), 956–966 (1991)

Winkler, E.: Die Lehre Von Elasticitaet Und Festigkeit, 1st edn. H. Dominicus, Prague (1867)

Wu, G., Finn, W.D.L.: Dynamic elastic analysis of pile foundations using finite element method in the frequency domain. Can. Geotech. J. **34**, 34–43 (1997a)

Wu, G., Finn, W.D.L.: Dynamic nonlinear analysis of pile foundations using finite element method in the time domain. Can. Geotech. J. **34**, 44–52 (1997b)

Zeevart, L.: Foundation Engineering for Difficult Sub-Soil Conditions, 2nd edn. Van Nostrand Reinhold Company, New York (1972)

Zienkiewicz, O.C., Taylor, R.L.: The Finite Element Method. Butterworth-Heinemann, Oxford (2005). ISBN 978-0-7506-6431-8

Zienkiewicz, O.C., Best, B., Dullage, C., Stagg, K.G.: Analysis of nonlinear problems in rock mechanics with particular reference to jointed rock systems. Supplemental Notes, vol. 3(8–14), pp. 501–509 Report/Paper Numbers: No. 364 (1970)

Behaviour of Laterally Loaded Piles in Soft Clay on Sloping Ground

Deendayal Rathod[1(✉)], K. Muthukkumaran[1], and T.G. Sitharam[2]

[1] Department of Civil Engineering, National Institute of Technology, Tiruchirappalli 620015, India
deendayalrathod123@gmail.com, kmk@nitt.edu
[2] Department of Civil Engineering, Indian Institute of Science, Bangalore 560012, India
sitharam@civil.iisc.ernrt.in

Abstract. This paper presents the results of static lateral load test carried out on single aluminum model pile embedded in soft clay (consistency index Ic = 0.42) on sloping ground. A series of laboratory model test had been carried out on the instrumented aluminum model pile on sloping ground of varying slopes (1V:1H, 1V:3H and 1V:5H) and with varying embedment length to diameter ratio (L/D) of 20, 25 and 30. From the experimental studies, it was observed that when ground surface changes from horizontal to steep slope (1V:1H), the reduction in pile capacity for L/D ratios 20, 25 and 30 was nearly 50%, 46% and 36% respectively, where as for ground surface changes from horizontal to 1V:5H slope, the reduction in pile capacity was observed to be 1–2% for all L/D ratios (20, 25 and 30).

1 Introduction

Piles have been extensively used for supporting axial and lateral loads for variety of structures including high rise buildings, transmission towers, power stations, offshore structures and highway and railway structures. Pile foundations are subjected to lateral loading due to an earth pressure, wind action, wave action, earthquake and impact of ships in dock and harbour structures. In case of coastal and offshore structures, the lateral loads are significantly high and the structure must withstand the berthing and mooring of ships. Lateral loads are in the order of 10–15% of the vertical loads in case of onshore structures and 25–30% in case of coastal and offshore structures (Rao et al. 1998). It is found that at various sites in the coastal region of India, the top layer is found to be soft clay or loose sand with a thickness that may vary from 5.0 m to 30.0 m or more (Boominathan and Ayothiraman 2007). Most of the structures which are constructed on soft clay with sloped ground surface are subjected to high lateral loads. For example, many high-rise buildings, bridges and transmission towers which are constructed on sloping ground in Hong Kong city are subjected to significant amount of lateral loads (Ng et al. 2001). The design of pile foundation in coastal region to resist lateral load is primarily based on the limiting deflection criterion, considering the safe operation of the superstructure. The design of offshore structures depends on various parameters such as wind forces and the loads from waves and currents. In such cases,

T. Abdoun and S. Elfass (eds.), *Soil Dynamics and Soil-Structure Interaction for Resilient Infrastructure*, Sustainable Civil Infrastructures, DOI 10.1007/978-3-319-63543-9_13

studying the interaction between the soil and the pile foundation due to lateral load is essential. Under lateral loads, the behaviour of pile placed in the sloping ground is different from piles placed in the horizontal ground, resulting in a reduction of pile carrying capacity. The reduction in pile capacity depends mainly on reduction of soil mass around the pile and the fixity condition of the pile. An early research on single pile shows that the ultimate capacity of pile was estimated assuming the deformations would be acceptable if an adequate factor of safety against ultimate failure was used to determine the allowable load capacity. The work of Matlock and Reese (1960) can be considered as one of the first attempts to understand laterally loaded pile behaviour. They have developed the solutions for soil reaction, bending moments and deflections for both short and long piles. Broms (1964) developed a solution for the ultimate lateral resistance of the pile assuming the distribution of lateral pile soil pressure and considering the statics of the problem. Poulos and Davis (1980) illustrated the influence of the pile soil system's relative stiffness on the behaviour of a single pile in terms of a relative stiffness factor $(K_{rc}) = E_p I_p / E_s L^4$ where E_p = modulus of elasticity of pile material; I_p = moment of inertia of pile section; E_s = average soil modulus; and L = depth of embedment of pile. The pile is said to be rigid if K_{rc} is greater than 10^{-2} and flexible if K_{rc} is less than 10^{-2}. Rajashree and sundaravadivelu (1996) reported analysis of laterally loaded pile in soft clay, idealizing the pile as beam element and the soil by non-linear inelastic spring element, modelled with elasto-plastic sub elements. An iterative procedure was adopted to perform a non-linear finite element analysis and the effect of static lateral load on load deflection behaviour was studied. Rao et al. (1996) carried out tests on model groups of piles to support dolphin-type structures. Static load tests have been conducted on instrumented model pile groups embedded in a marine clayey bed. The spacing between piles, number of piles in a group, and arrangement of pile group with respect to the direction of lateral loading have been varied. The results indicated that the capacity of pile group not only depends on the spacing between the piles, but also on the arrangement of piles in the group. Mezazigh and Levacher (1998) conducted the centrifuge test on model single aluminium pile placed on horizontal ground and sloping ground surface in cohesionless soil by varying the distance from the crest of slope. The effect of distance to the slope, slope angle and soil properties were studied. The load versus deflection curves; bending-moment and p-y curves were derived for piles close to slopes and compared to the horizontal ground response. Rao et al. (1998) conducted experiments on pile groups and studied the influence of parameters like flexural rigidity of pile material, embedment length of pile and arrangement of piles on the behaviour of laterally loaded pile groups. The results indicated that the lateral load capacity of the pile group depends mainly on the rigidity of pile soil system for different arrangements of piles within a group. Ng et al. (2001) studied the influence of laterally loaded sleeved piles and pile groups of slope stability. The stability of the slope was evaluated using the strength reduction technique. The evolution of slope failure is examined and the factors of safety for both initiation of instability and global failure of the slope are identified from the numerical analyses. It is found that the sleeving technique is capable of reducing the stresses in the shallow depths of the slope in front of the piles significantly, thus improving the local stability of the slope, but offers limited benefit with respect to global stability. Chae et al. (2004)

performed several numerical studies with a 3D FE model test and prototype test on laterally loaded short rigid piles and pier foundation located near a 30 degrees slope. The lateral resistance of pile was observed to be decreasing with the change in location closer to the crest of the slope. Ilyas et al. (2004) conducted a series of centrifuge model tests to examine the behavior of laterally loaded pile groups of 2, 2×2, 2×3, 3×3, and 4×4 piles with a center-to-center spacing of three or five times the pile width in normally consolidated and over consolidated kaolin clay. It is established that the pile group efficiency reduces significantly with increasing number of piles in a group due shadowing effect. And also, it is found that the front piles experience larger load and bending moment than that of the trailing piles. Boominathan and Ayothiraman (2007) carried out the static and dynamic lateral load tests on model aluminium single piles embedded in soft clay to study their bending behaviour. The results indicated that the maximum dynamic bending moment of pile in soft clay is about 1.5 times higher than the maximum static bending moment. Muthukkumaran et al. (2008) conducted the laboratory experiments on sand with different relative densities to study the effect of slope angle on lateral load capacity of pile and pile groups and studied the effect of relative density on bending moment, lateral soil resistance and lateral deflection due to surcharge load and finally developed p-y design curve in non-dimensional form. Chandrasekaran et al. (2010) carried out the static lateral load tests on $1 \times 2, 2 \times 2$, 1×4, and 3×3 aluminium model pile groups embedded in soft clay. The effects of pile spacing, number of piles, embedment length, and configuration on pile-group interaction were investigated. From the experimental results, it has been found that the lateral capacity of piles in 3×3 group at three diameter spacing is about 40% less than that of the single pile and causes 20% increases in the maximum bending moment when compared with single pile. Georgiadis and Georgiadis (2010) carried out three-dimensional finite element analyses to study the behavior of piles embedded in cohesive sloped ground under undrained lateral loading conditions. Piles of different diameter and length in slope were considered. From the analysis, analytical formulations are derived for the ultimate load per unit length and the initial stiffness of hyperbolic p-y curves. New p-y criteria for static loading of piles in clay are proposed, which takes into account the inclination of the slope and the adhesion of the pile-slope interface. Kim et al. (2011) conducted a series of field and laboratory load tests on marine clay to determine the pile behavior. They have proposed a new hyperbolic p-y curve with an improved wedge model which is applicable for large-diameter laterally loaded piles embedded in clayey soil. Rajashree and Sitharam (2001) developed a finite-element model of batter piles under lateral loads and performed static and cyclic load analyses adopting an incremental-iterative procedure where the pile is idealized as beam elements and the soil as elasto-plastic spring elements. The proposed numerical model highlighted the importance of the degradation factor and its influence on the soil resistance-displacement (p-y) curve, number of cycles of loading, and cyclic load response. Georgiadis and Georgiadis (2012) carried out three-dimensional finite element analyses using Plaxis 3D Foundation and presented the response of laterally loaded piles of different geometries, installed at several distances from slopes of various inclinations. Also, developed p-y curves for the case of undrained lateral loading of piles near the crest of clay slopes and implemented the p-y curve into a commercial subgrade reaction computer code to perform a series of parametric numerical analyses.

Muthukumaran (2014) conducted the laboratory model tests to study the effect of slope and loading direction on laterally loaded piles in cohesionless soil. From the experimental investigation, it is observed that if the pile is placed 15 times the pile diameter away from the slope crest within the embankment, the influence of slope in the lateral-load capacity is almost negligible under both forward and reverse lateral load. Madhusudan Reddy and Ayothiraman (2015) carried out 1-g model experiments in the laboratory to investigate the behavior of a single pile in sand under combined uplift and lateral load. The aluminum model pile having outer and inner diameters of 25.4 and 19.0 mm, respectively was used. Different length-to-diameter ratios of 18, 28, and 38 were considered by varying the pile length to simulate behavior of both stiff and flexible piles. Zhang et al. (2015) set up four hypothetical cases related to laterally loaded piles in uniform and layered soil systems to discuss the response of the laterally loaded pile in layered soils. The comparison results indicate that the pile behavior was controlled by the sub-grade soil stiffness at shallow depth (up to 3–4 times the pile diameter).

From the literature it is clear that only very few limited research works have been carried out on piles subjected to lateral load in sloping ground (Mezazigh and Levacher 1998; Ng et al. 2001; Georgiadis and Georgiadis 2010; Muthukkumaran et al. 2008; Sawant and Shukla 2013; Muthukkumaran 2014) and the behaviour of pile embedded on crest of soft clay requires further study. Therefore, this paper aims to fill this gap through a comprehensive experimental investigation carried out on single piles embedded in soft clay under static lateral load. The main objective of the present investigation is to study the behaviour of long flexible single pile located on crest of sloping ground of soft clay with varying slopes (1V:1H, 1V:3H and 1V:5H) and embedment length to diameter ratios (L/D 20, 25 and 30) under lateral load.

2 Experimental Set-up

2.1 Testing Tank and Model Pile

The dimensions of the test tank were decided based on influence zone of soil mass from pile. The influence zone of soil mass from pile was observed to be 10 times the pile diameter in the direction of loading for piles under static lateral load (Poulos and Davis 1980; Rao et al. 1998). Accordingly, the tank size of 2 m × 1 m × 1 m was chosen to perform the static lateral load tests. The static lateral load was applied by means of dead weights placed on a hanger connected to a flexible steel wire, strung over a frictionless pulley supported by a loading platform as shown in Fig. 1. An aluminium pipe pile having a length of 810 mm and outer diameter of 25.46 mm with 1 mm wall thickness was used as a model pile. The modulus of elasticity ($E = 5.0117 \times 10^4$ N/mm^2) and flexural rigidity of the pile ($EI = 2.885 \times 10^8$ N-mm^2) were obtained by conducting simple bending test. To get the accurate results, different loading cases were considered. The moment of inertia of the pile (I) was estimated as 5756.45 mm^4 and Poisson's ratio (μ) as 0.3. Based on the relative stiffness factor *(K_{rc})*, the piles were classified as rigid or flexible (Poulos and Davis (1980)). The K_{rc} given as, $K_{rc} = E_p I_p / E_s L^4$. The pile is said to be rigid if $K_{rc} > 10^{-2}$ and flexible if $K_{rc} < 10^{-2}$.

For the selected soil and pile properties, the pile having length (*L*) to diameter (*D*) ratio of 20, 25 and 30 behaves as long flexible. The model pile (aluminium pipe) was instrumented with electrical resistance type strain gauges having gauge length of 3 mm, gauge factor (K) of 2 and resistance (R) of 120 ohms. The strain gauges were pasted on one side of pile with its axis parallel to the pile axis forming full bridge. The strain gauges were placed at a minimum spacing of 50 mm apart with bottom gauge placed 50 mm above the pile toe and top gauge placed 60 mm below the pile head. Small holes were made on the side of the pile slightly above each gauge location to pass the wires inside the pile. The strain gauges were protected against direct contact with soil by applying a thin layer of silicon gel over the strain gauges. After the placement of silicone gel, the resistance of the strain gauges were checked to ensure the workability of the strain gauges. The bottom of the pile was closed by a 60° conical shoe to facilitate easy installation of piles and to prevent soil plugging into the hollow model piles. The test pile and the location of strain gauges are shown in Fig. 2. Strain gauges were calibrated for bending moment along the pile length by conducting a simple bending test considering simply supported ends. The strain response was found to be linear with the bending moment for the applied load for all the strain gauges pasted in the pile. The calibration constant was obtained from the slope of the straight line. An average value of linear response of all the train gauges has been taken as bending constant and found to be 30.453 N-mm per micro strain. The bending moment M for each strain value can be obtained by using Eq. (1)

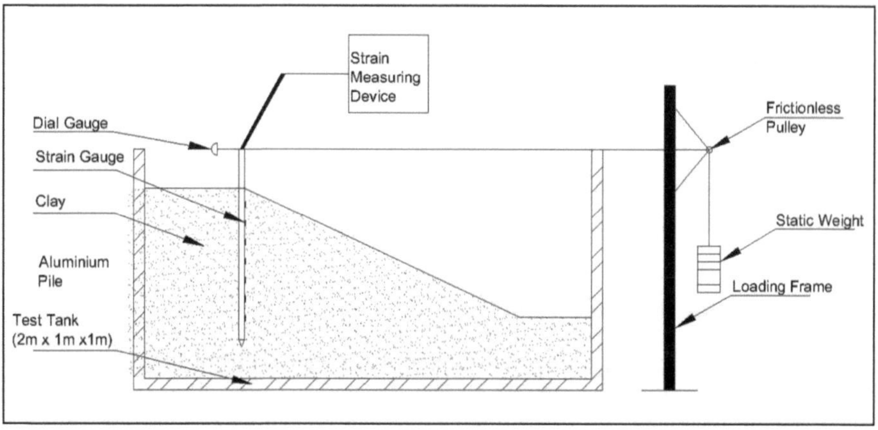

Fig. 1. Experimental set-up

$$M = C * \varepsilon \tag{1}$$

Where

C – bending constant (1 micro strain = 30.453 N-mm)
ε – Strain (micro strain).

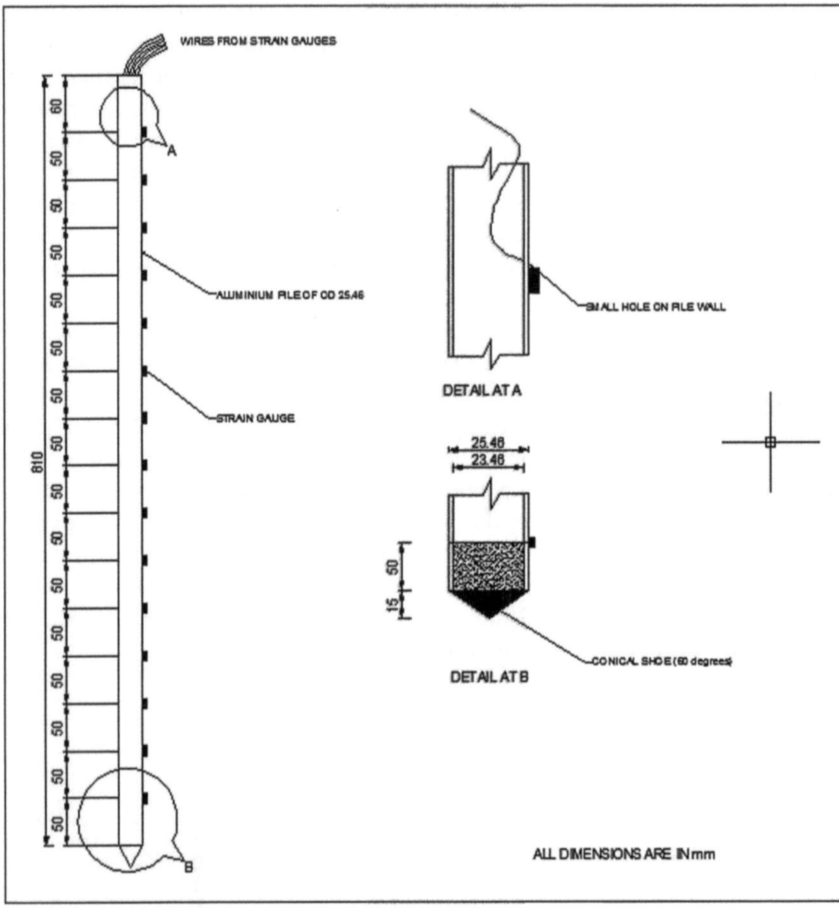

Fig. 2. Pile instrumentation and location of strain gauges

2.2 Soil Used for the Present Study

For the experimental studies, clay soil sample was collected from the coastal area at Nagapattinam, Tamil Nadu (INDIA). The laboratory tests like sieve analysis, hydrometer analysis and Atterberg limits were carried out to classify the soil. The undrained shear strength of the clay was also determined by conducting the laboratory vane shear test. The complete physical properties of the soil are presented in Table 1. The soil is classified as fat clay (CH) as per ASTM D2487-06.

2.3 Test Procedures

In the present study, the clay bed was prepared similar to the procedure adopted by Rao et al. (1998) and Chandrasekaran et al. (2010). The clay was mixed in a separate mixing tank with the required amount of water to get the soft consistency (I_c = 0.42) of

Table 1. Physical properties of soil

Properties	Values
Specific Gravity (G)	2.61
Liquid Limit (LL)	75%
Plastic Limit (PL)	35%
Shrinkage Limit (SL)	17%
Unit Weight of soil (γ)	17.8 kN/m^3
Undrained Strength (S_u)	N/m^2

a typical clay deposit in Nagapattinam region and kept for two days for proper mixing. The model pile was placed in the centre of the test tank using templates and then the uniformly mixed clay was placed and hand packed in the test tank in layers of 50 mm thickness. Each layer was tamped with a template to remove the entrapped air to ensure homogenous condition. In this case, the installation procedure simulates the bored pile condition. The water content was determined on the soil samples collected from various depths around the tank as the same procedure adopted by Boominathan and Ayothiraman (2007). The water content of soil samples varied between 57.4 to 59.2% and undrained shear strength which was measured by vane shear apparatus at the above mentioned water content was found to be between 9.8 and 10.2 kN/m^2, which indicates that the clay is nearly homogeneous in the test bed.

2.4 Tests Carried Out

A series of static lateral load tests had been carried out on the instrumented model pile on clayey soil of soft consistency as shown in Table 2. The static lateral load was applied in increments by adding weights in a loading arrangement as shown in Fig. 1. The load was applied at the pile head (60 mm above the top strain gauge). For all the L/D ratios, the loading point was same as the pile head. However, the depths of embedment were varied with respect to L/D ratios. For L/D 20, 25 and 30, the depth of embedment of piles were 508 mm, 635 mm and 762 mm respectively from the pile toe. The ultimate or allowable theoretical capacity was calculated for the given soil and pile size. The load test was performed in several loading stages. In each load increment, the load was applied 20% of the ultimate load and each load was kept for a settlement of 0.01 mm for 30 min (as per IS2911 Part IV). For each increment of loading, the horizontal deflection of the pile was measured at pile head using dial gauges. When the deflection of the pile ceases, the next load increment was applied. The lateral load was applied till the lateral deflection reaches 20% of pile diameter (5 mm) and the corresponding load were taken as allowable lateral load capacity of the pile (Rao et al. 1998; Chandrasekaran et al. 2010). The strain gauge readings were also measured along the length of pile for each increment of load using Data Acquisition System (DAS). The vertical settlement of pile was also measured by the dial gauge for few tests and was found to be negligible. Further it is assumed that the applied load is purely in horizontal direction.

Table 2. List of experiments conducted

Soil slope	Embedment length (mm)	L/D ratio
Horizontal	509.2	20
	636.5	25
	763.8	30
1V:5H	509.2	20
	636.5	25
	763.8	30
1V:3H	509.2	20
	636.5	25
	763.8	30
1V:1H	509.2	20
	636.5	25
	763.8	30

3 Results and Discussion

3.1 Lateral Load Versus Pile Head Deflection Behavior of a Single Model Pile

Figures 3, 4 and 5 shows the lateral load versus pile head deflection curves for L/D ratios 20, 25 and 30 respectively for different ground slopes. When ground surface changes from horizontal to 1V:1H or 1V:3H or 1V:5H slopes, the lateral load capacity reduces significantly for all the L/D ratios (20, 25 and 30). This reduction in pile capacity is due to the reduction in passive resistance of soil in front of the pile due to slopes. When ground surface changes from horizontal to steep slope (1V:1H), the reduction in pile capacity for L/D ratios 20, 25 and 30 is nearly 50%, 46% and 36% respectively, where as for ground surface changes from horizontal to 1V:5H slope, the reduction in pile capacity is just 1–2% for all L/D ratios (20, 25 and 30). It is also observed that the behaviour of pile is significantly changes up to ground slope 1V:3H. When ground surface changes from horizontal to 1V:5H slope, the behaviour of pile is almost like horizontal ground surface. It shows very clearly that if slope angle is less than 11.30° (1V:5H), the effect of slope is almost negligible on lateral load pile capacity of single pile. This is noticed for all L/D ratios 20, 25 and 30. It is also observed that the behaviour lateral load-deflection curves for L/D 25 and 30 (Figs. 4 and 5) is almost same up to 0.5 mm deflection for all slopes, which is almost 2% of the pile diameter (D). This implies that the effect of slope is negligible for smaller lateral loads. The load carrying capacity of pile having L/D 25 for all slopes at 0.5 mm deflection is 25 N, where as for L/D 30, the load carrying capacity is 50 N. This increase in pile capacity is due to increase in the relative stiffness of the pile and the soil system.

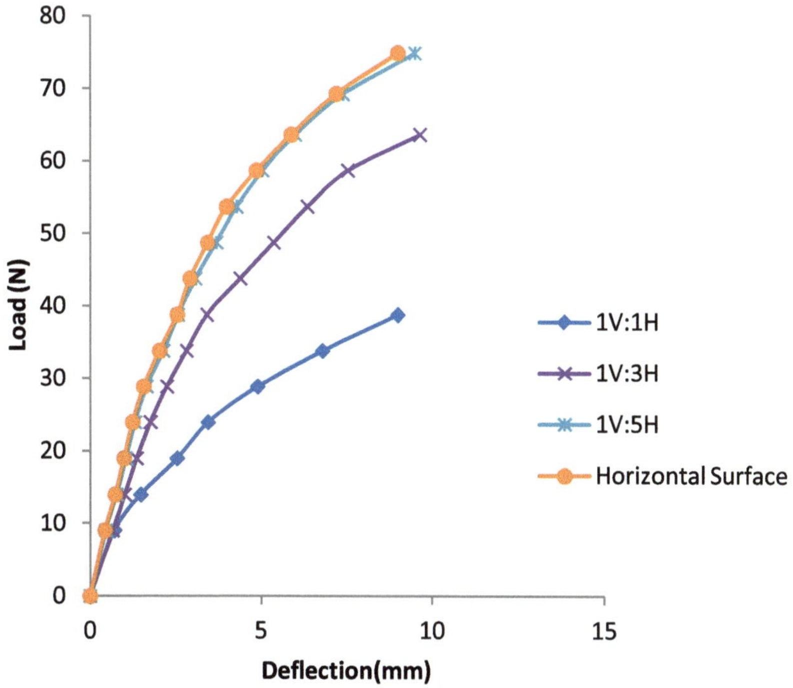

Fig. 3. Lateral load - deflection curves for different slopes (L/D 20)

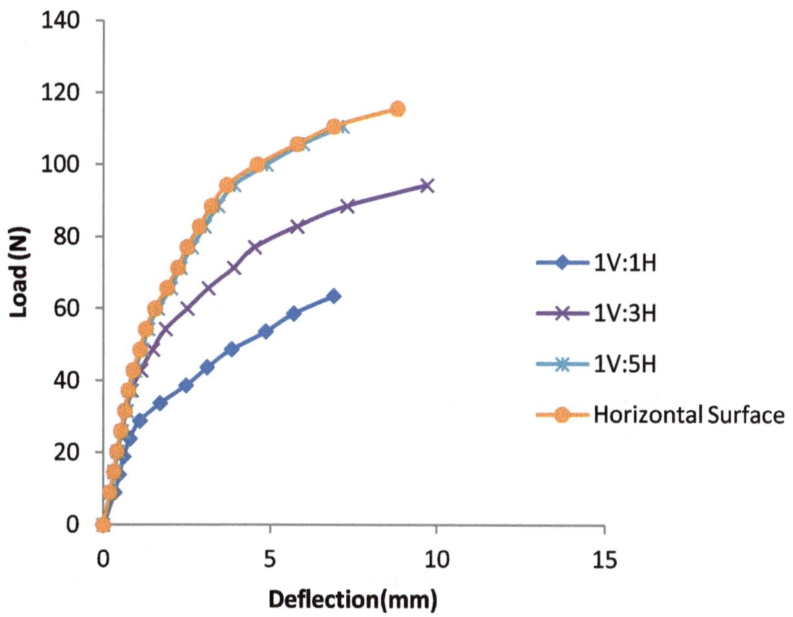

Fig. 4. Lateral load - deflection curves for different slopes (L/D 25)

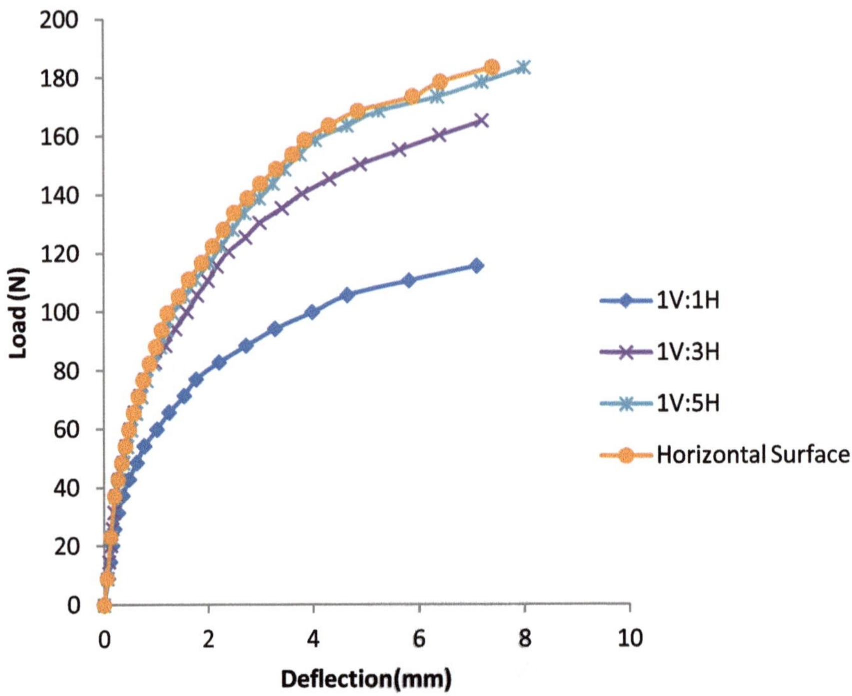

Fig. 5. Lateral load - deflection curves for different slopes (L/D 30)

3.2 Bending Moment Behaviour of a Single Model Pile in Horizontal Ground Surface

Figures 6, 7 and 8 shows the bending moment variation along the depth of pile for horizontal ground surface for different L/D ratios (20, 25 and 30). From these figures, it is found that the maximum bending moment increases with increase in applied loads. It is also observed that the increase in L/D ratio significantly changes the bending moment variation along the depth of the pile. As L/D ratio increases, the bending moment decreases. This is due to increase in the relative stiffness of the pile-soil system and also due to increase in the embedded length of the pile. This also shows that the increase in L/D ratio increases the long flexible behaviour of the pile.

3.3 Bending Moment Behaviour of a Single Model Pile in Sloping Ground Surface

Figures 9, 10 and 11 shows the typical bending moment variation along the depth of the pile for 1V:3H slope for different L/D ratios (L/D 20, 25 and 30). From these figures, it is found that the change in ground surface from horizontal to slopes affects the bending behaviour of the pile significantly. The bending moment increases with increase in applied loads. And also, the bending moment increases with increase in slopes. However, this increment is very high for piles having lesser L/D ratio (20) and

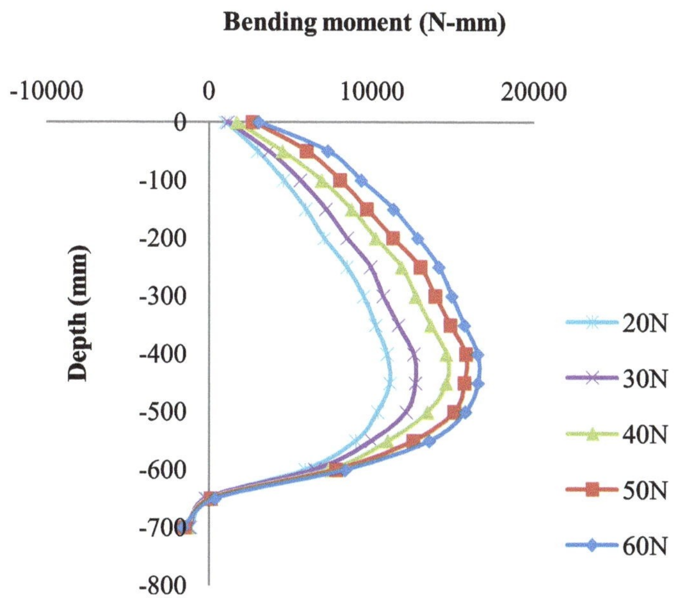

Fig. 6. Variation of bending moments along the depth of pile for L/D 20, horizontal ground surface

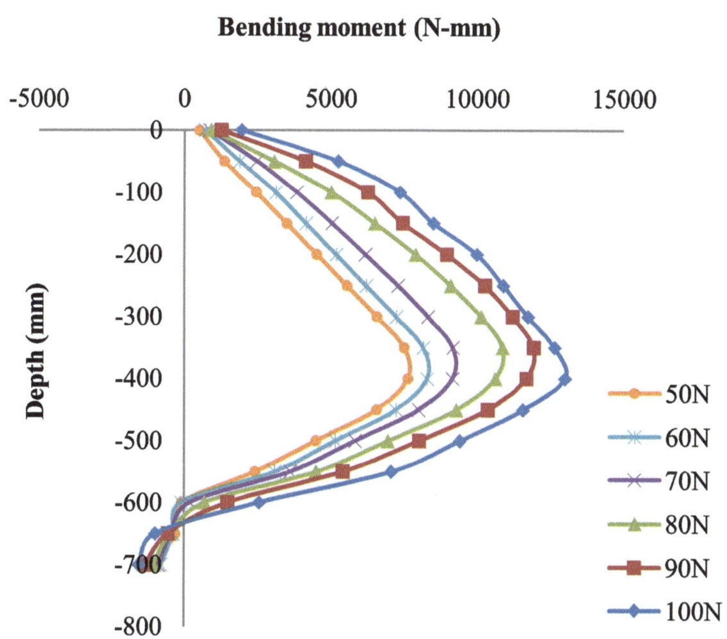

Fig. 7. Variation of bending moments along the depth of pile for L/D 25, horizontal ground surface

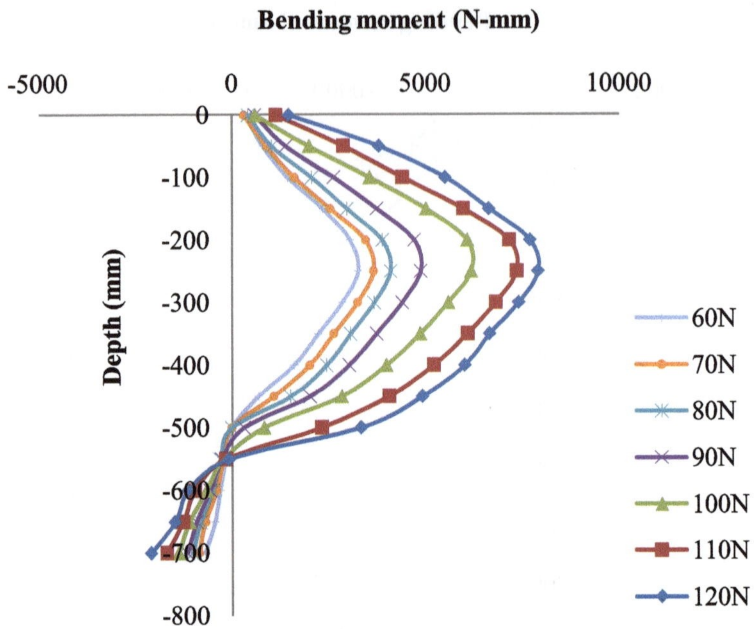

Fig. 8. Variation of bending moments along the depth of pile for L/D 30, horizontal ground surface

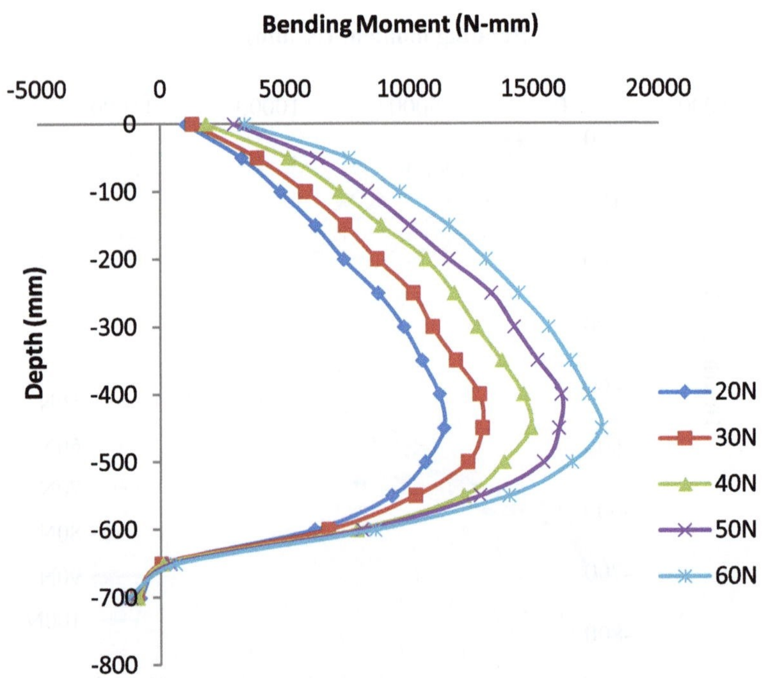

Fig. 9. Variation of bending moments along the depth of pile for L/D 20, 1V:3H slope

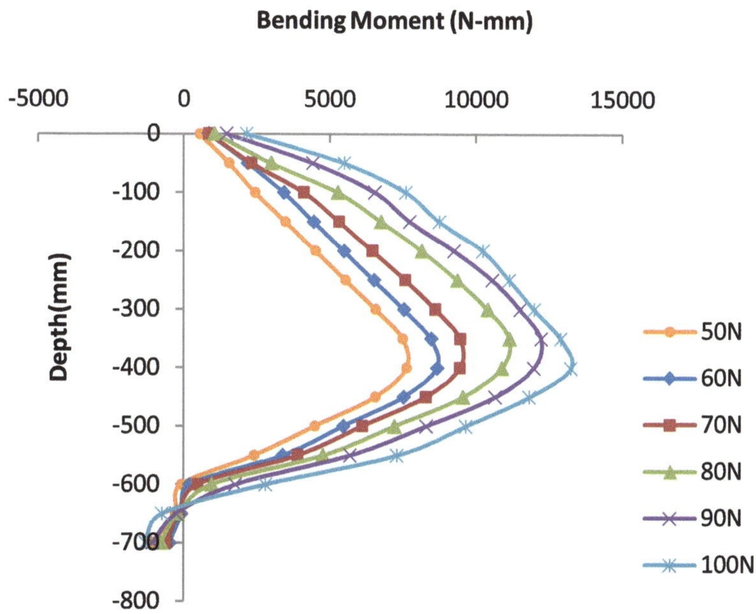

Fig. 10. Variation of bending moments along the depth of pile for L/D 25, 1V:3H slope

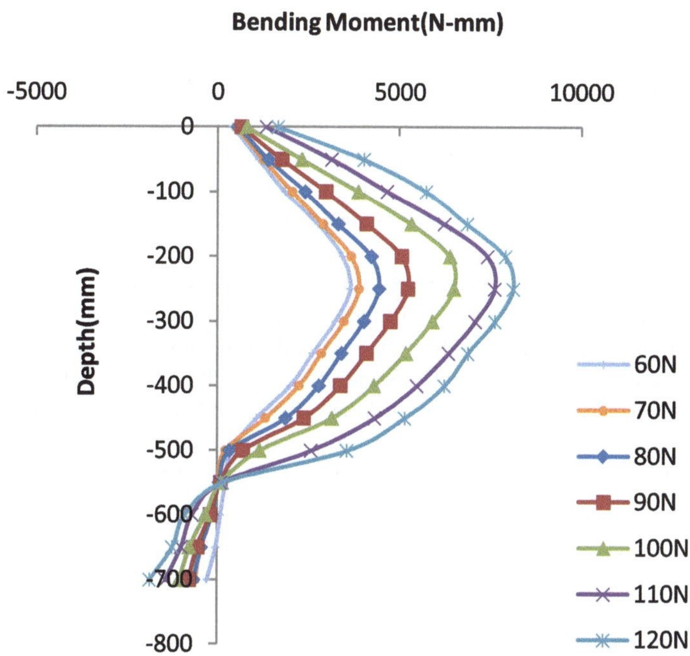

Fig. 11. Variation of bending moments along the depth of pile for L/D 30, 1V:3H slope

steeper slope (1V:1H). This is due to the decrease in the resistance in front of pile as there is a reduction in the soil mass due to sloping ground. It is also observed that the depth at which the maximum bending moment occurs (Depth of fixity), decreases with increase in the L/D ratios. This is due to increase in the relative stiffness of the pile and the soil due to increase in the embedded length of the pile.

4 Conclusions

The results obtained from the experimental studies carried out on a single model aluminium pile embedded on soft clay with sloping ground are presented and discussed.

- It is observed that when ground surface changes from horizontal to 1V:1H or 1V:3H or 1V:5H slopes; the lateral load capacity reduces significantly for all the L/D ratios 20, 25 and 30. This reduction in lateral load capacity is due to the reduction in passive resistance of the soil in front of the pile. However, the rate of change of reduction in lateral loads capacity is much less when ground surface horizontal ground to 1V:5H slope than 1V:3H or 1V:1H slopes.
- When ground surface changes from horizontal to steep slope (1V:1H), the reduction in pile capacity for L/D ratios 20, 25 and 30 is nearly 50%, 46% and 36% respectively, where as for ground surface changes from horizontal to 1V:5H slope, the reduction in pile capacity is observed to be 1–2% for all L/D ratios (20, 25 and 30). From the study, it shows very clearly that if slope angle is less than 11.30° (1V:5H), the effect of slope is almost negligible on lateral load pile capacity of a single pile.
- It is also observed the increase in L/D ratio increases the lateral load capacity. And the horizontal surface carries more loads as compare to sloping ground. From the results it is clearly observed that the increase in L/D ratio increases the lateral load capacity even though the ground surface is sloped. So, it can be concluded that the sloping ground in front of pile plays an important role in clayey soil and causes reduction in pile capacity.
- It is observed that the increase in L/D ratio decreases the bending moments. This is due to increase in the relative stiffness of the soil-pile system and also due to increase in the embedded length of the pile. This also shows that the increase in L/D ratio increases the long flexible behaviour of the pile. It is also found that the change in ground surface from horizontal to slopes affects the bending behaviour of the pile significantly. The bending moments increases with increase in slopes. However, this increment is very high for piles having lesser L/D ratio (20) and with steeper slope (1V:1H).

References

Boominathan, A., Ayothiraman, R.: An experimental study on static and dynamic bending behavior of piles in soft clay. Geotech. Geol. Eng. **25**, 177–189 (2007)

Boominathan, A., Ayothiraman, R.: Dynamic response of laterally loaded piles in clay. Proc. Inst. Civ. Eng. **4**, 233–241 (2005)

Broms, B.B.: Lateral resistance of piles in cohesive soils. J. Soils Mech. Found. Div. **90**(2), 27–63 (1964). ASCE

Chandrasekaran, S.S., Boominathan, A., Dodagoudar, G.R.: Group interaction effects on laterally loaded piles in clay. J. Geotech. Geoenviron. Eng. **136**(4), 573–582 (2010)

Chae, K.S., Ugai, K., Wakai, A.: Lateral resistance of short single piles and pile groups located near slopes. Int. J. Geomech. **4**(2), 93–103 (2004)

Georgiadis, K., Georgiadis, M.: Undrained lateral pile response in sloping ground. J. Geotech. Geoenviron. Eng. **136**(11), 489–1500 (2010)

Georgiadis, K., Georgiadis, M.: Development of p-y curves for undrained response of piles near slopes. Comput. Geotech. **40**, 53–61 (2012)

Ilyas, T., Leung, C.F., Chow, Y.K., Budi, S.S.: Centrifuge model study of laterally loaded pile groups in clay. J. Geotech. Geoenviron. Eng. **130**(3), 274–283 (2004)

Madhusudan Reddy, K., Ayothiraman, R.: Experimental studies on behavior of single pile under combined uplift and lateral loading. J. Geotech. Geoenviron. Eng. **141**(7), 04015030 (2015). ASCE

Matlock, H., Reese, L.C.: Generalized solutions for laterally loaded piles. J. Soil Mech. Found. Div. **86**(5), 63–91 (1960). ASCE

Mezazigh, S., Levacher, D.: Laterally loaded piles in sand: slope effect on P-Y reaction curves. Can. Geotech. J. **35**(3), 433–441 (1998)

Muthukkumaran, K., Sundaravadivelu, R., Gandhi, S.R.: Effect of slope on p-y curves due to surcharge load. Soils Found. **48**(3), 353–361 (2008)

Muthukkumaran, K.: Effect of slope and loading direction on laterally loaded piles in cohesionless soil. Int. J. Geomech. **14**(1), 1–7 (2014). ASCE

Rao, S.N., Ramakrishna, V.G.S.T., Raju, G.B.: Behavior of pile-supported dolphins in marine clay under lateral loading. J. Geotech. Geoenviron. Eng. **122**(8), 607–612 (1996)

Rao, S.N., Ramakrishna, V.G.S.T., Rao, M.B.: Influence of rigidity on laterally loaded pile groups in marine clay. J. Geotech. Geoenviron. Eng. **124**(6), 542–549 (1998)

Ng, C.W.W., Zhang, L.M., Ho, K.K.S.: Influence of laterally loaded sleeved piles and pile groups on slope stability. Can. Geotech. J. **38**(3), 553–566 (2001)

Poulos, H.G., Davis, E.H.: Pile Foundation Analysis and Design. Wiley, New York (1980)

Rajashree, S.S., Sitharam, T.G.: Nonlinear finite-element modeling of batter piles under lateral load. J. Geotech. Geoenviron. Eng. **127**(7), 604–612 (2001). ASCE

Rajashree, S.S., Sundaravadivelu, R.: Degradation model for one-way cyclic lateral load on piles in soft clay. Comput. Geotech. **19**(4), 289–300 (1996)

Randolph, M.F.: The response of flexible piles to lateral loading. Geotechnique **31**, 247–259 (1981)

Kim, Y., Jeong, S., M.ASCE, Lee, S.: Wedge failure analysis of soil resistance on laterally loaded piles in clay. J. Geotech. Geoenviron. Eng. **137**(7), 678–694 (2011)

Zhang, L., Zhao, M., Zou, X.: Behavior of laterally loaded piles in multilayered soils. Int. J. Geomech. **15**(2), 06014017 (2015). ASCE

Micromechanical Modeling of the Seismic Response of Gravity Retaining Walls

Usama El Shamy(✉) and Aliaksei Patsevich

Civil and Environmental Engineering Department,
Southern Methodist University, Dallas, TX, USA
uelshamy@lyle.smu.edu, apatsevich@smu.edu

Abstract. In this study, the analysis of the seismic response of a soil-retaining wall system is performed based on a three-dimensional microscale framework utilizing the discrete element method (DEM). The proposed method is employed to investigate the seismic response of gravity-type retaining walls with three degrees of freedom. In the simulation, the granular soil deposit is idealized as a collection of spherical soil particles; the retaining wall is simulated as a rigid block composed of clumped particles to yield the physical characteristics of a real-life retaining wall. The model is processed under the gravitational acceleration of 50 g to reduce the total duration of the simulation and dimensions of the model.

The model accounts for the effects of nonlinear soil behavior and possible separation between the retaining wall and soil deposit. The impact of amplitude and frequency of input dynamic excitation on the response of the wall was analyzed under different conditions. The computational approach is able to capture essential dynamic response patterns such as the effect of resonance on amplifying the response and the failure of the wall as it underwent excessive rotation and displacement due to increased earth pressure and inertia forces. In addition, time histories of the resultant lateral thrust on the wall show that its magnitude varies as the wall moves relative to the backfill and its location changes as well. The shape of the failure wedge was also captured from analysis of maximum shear strains developed in the backfill. Under strong ground motion, wall failure was associated with seismic bearing capacity failure near the wall toe due to increased wall pressure.

1 Introduction

Earth retaining structures, such as retaining walls, are commonly used in seismically active areas. Several historical earthquakes have resulted in significant permanent deformation and failure of retaining structures. Seismic analysis of retaining walls is a challenge as wall movements and pressures depend on several factors such as the response of the underlying soil, the response of the backfill, inertial forces of the wall, and the characteristics of the input motion. Only a few well-documented case histories involving field measurements of wall response are available, thus most of the current understanding of the dynamic response of retaining walls has come from model tests and numerical analyses.

T. Abdoun and S. Elfass (eds.), *Soil Dynamics and Soil-Structure Interaction for Resilient Infrastructure*, Sustainable Civil Infrastructures, DOI 10.1007/978-3-319-63543-9_14

Various simplified models have been used in order to determine effects of the earthquake on the retaining structures. Okabe (1926), Mononobe and Matsuo (1929) developed the basis of pseudo-static analysis (known as the M-O method) of seismic earth pressures on retaining structures. Further development on the Mononobe-Okabe method has been done by Richards and Elms (1979), Whitman and Liao (1985) when they employed Newmark's sliding block procedure in evaluations of earthquake induced displacements of gravity retaining walls. Despite all the improvements and modifications to the Newmark's sliding block method, the rigid block approach lacks the ability of modeling the seismic response of the backfill behind a retaining wall and, consequently, the associated effects on earthquake-induced displacements and dynamic wall thrust (Nadim and Whitman 1983; Kramer and Smith Kramer 1997; Wartman et al. 2003).

Design based on pseudo-static approach is generally considered conservative, since even when the safety factor drops below one the soil structure could experience only finite displacement rather than a complete failure. Analysis techniques used to determine the response of retaining walls to the dynamic loading typically implement multiple assumptions and simplifications limiting the application of the solution to a specific case that is being investigated. In this paper, a new approach for investigating the complex behavior of retaining walls under a seismic excitation using the Discrete Element Method (DEM) is presented. The proposed approach accounts for soil nonlinearity, dynamic soil-wall interaction, and possible separation between the wall and the surrounding soil during dynamic excitation.

2 Model Description

A microscale-based approach is adopted in this study to model the soil-retaining wall system. The soil deposit is modeled as a collection of granular particles using DEM, resting on a rigid base. The motion of a discrete particle is dictated by the translational and angular momentum equations of motion. The retaining wall is idealized as a rigid body composed of clumped particles to simulate the physical characteristics of a real-life retaining wall. That is, regardless of the forces acting upon it, the block will not break apart. The contact forces between the clumped particles are not accounted for during the DEM calculation cycle. However, the interactions between the clumped particles and soil particles composing the deposit are considered. Since the clumped particles behave as a rigid body, the translational and rotational equations of motion are sufficient to describe its motion. More details of employed model maybe found in Zamani and El Shamy (2012), Patsevich (2015).

3 Computational Simulation

The proposed approach was employed to investigate the response of a three degrees of freedom retaining wall on a dry granular deposit. Figure 1 shows the deposit used in the simulation. Use was made of the high g-level concept commonly used in centrifuge testing to reduce the dimensions of the domain that needed to be filled with particles and to benefit from the shorter time scaling law within which the simulation can be conducted (Iai et al. 2005). Furthermore, periodic boundaries were employed in the lateral direction parallel to the direction of shaking to reduce the size of the simulation. The other two lateral boundaries and the base of the deposit were modeled as a rigid wall. The two walls in a lateral direction were created at a significant distance away from the retaining wall (18 m from the back of the wall and 6 m in front of the wall) in order to decrease any possible effects of the lateral boundaries. After the boundaries of the domain were established, the soil-wall system was created within the domain.

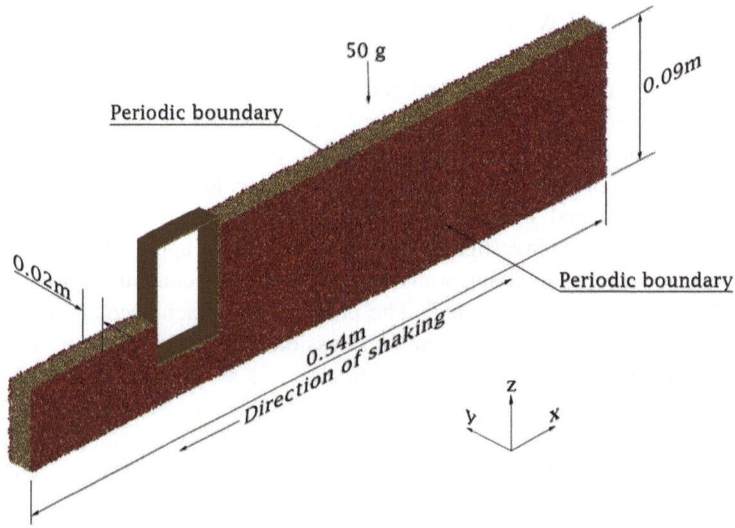

Fig. 1. Granular deposit and retaining wall structure as modeled in DEM simulations

Soil particles constituting the domain were generated and allowed to settle under gravity. New particles constituting the retaining wall were created and clumped together in a fashion that will keep the particles together no matter what force is acting on them. The retaining wall is composed of four walls to represent a box with a width of 0.02 m, a length of 0.06 m and a height of 0.09 m (width = 1 m, length = 3 m, height = 4.5 m in prototype units). These walls were made of clumped particles with a diameter of 2 mm and a center-to-center spacing of 0.5 mm (particles overlap to provide a relatively flat surface for the walls). The angle of friction between the wall and the surrounding soil was assumed to be 18°. The density of the concrete consisting retaining wall in prototype units was assumed to be 2400 kg/m^3, which resulted in each particle composing the wall to have a density of 2960 kg/m^3, and the total mass of the retaining wall to be 0.26 kg (32,500 kg in prototype units).The retaining wall was then installed on top of the base layer and more particles were generated to form the backfill. Wall stability checks were performed under static conditions. Computational data are summarized in Table 1 along with the computed static safety factors for wall stability. Additional computational details and information about model generation maybe found in Patsevich (2015).

Table 1. Simulation data in model units.

Soil deposit	
Number of particles	379,735
Particle diameter	1.2 to 1.8 mm
Density of solid particles	2650 kg/m^3
Particle normal/shear stiffness	5×10^5 N/m
Friction coefficient	0.5
Shear wave velocity	133 m/s
Average friction angle ϕ	27°
Retaining wall	
Length	60 mm
Width	20 mm
Height	90 mm
Static factors of safety	
Bearing capacity	4.6
Sliding	6.2
Overturning	2.1

The system was then subjected to a dynamic excitation applied to the bedrock and lateral boundaries perpendicular to the direction of shaking, and the responses of soil deposit and retaining wall were monitored. This application of the input motion is somewhat similar to shaking an experimental model retaining wall in a rigid box. The dynamic excitation followed a sinusoidal pattern with a frequency of 2 Hz (in prototype units) and consisted of three stages that lasted a total of eight seconds (in prototype units). During the first three seconds of the simulation the amplitude of the input motion was gradually increasing to reach the maximum amplitude of 0.1 g (in prototype units) at 3 s, then for the next four seconds the amplitude of the input motion stayed constant, and finally within one second the amplitude of the input motion gradually decreased to zero. The location of the rigid lateral boundaries was far enough to produce free-field behavior half way between the wall and the right boundary. This was verified by computing the amplification factor of the input motion as it propagates to the surface and comparing it to that from classical wave propagation theory. The computed one was about 1.35 compared to the theoretical value of 1.2.

Figure 2 shows the time-history of the acceleration of the input motion compared to the acceleration of the center of gravity of the wall. The response of the retaining wall is amplified relative to the input motion on the negative side of the acceleration record, i.e., when wall is moving away from the backfill. It can be noted that the inertial forces of the retaining wall have resulted in a phase lag between the motion of the retaining wall and the input motion applied to the bedrock. The maximum acceleration of the wall under current loading parameters was determined to be 0.15 g, which is 1.5 times larger than the amplitude of the input motion.

Time-histories for total soil thrust on front and back faces of the wall during the simulation (Fig. 3) show how the total thrust was changing throughout the simulation. According to the theoretical calculations based on Coulomb earth pressure theory,

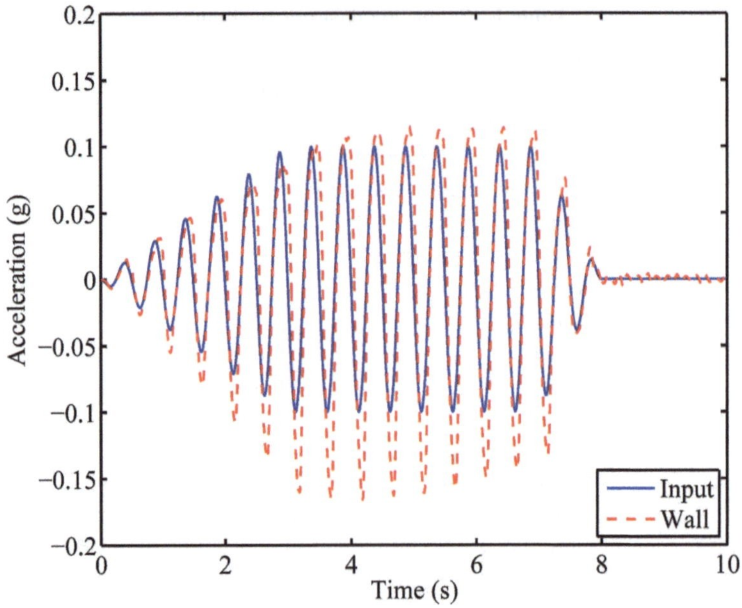

Fig. 2. Time histories of wall and input accelerations

minimum active thrust resulting from the backfill is 47.4 kN/m, while the computed initial thrust on the back of the wall is approximately 65.2 kN/m. Dynamic active soil thrust was calculated to be about 57.8 kN/m using the M-O method (using the characteristics of the input motion and a kh value of 0.1). The total thrust on the back side of the wall during the simulation is varying between 62 kN/m and 84 kN/m (Fig. 3). Values of the backfill thrust acting on the wall at the time instances when the wall is moving away from the backfill, i.e., lowest values of the total thrust, are comparable to the theoretical value of active earthquake pressure calculated using the M-O method. It can indicate that the wall experienced movement that resulted in the development of an active wedge when the wall was moving away from the backfill during the simulation. Comparatively, when the wall was moving towards the backfill, i.e., highest values of the total thrust, it did not produce enough movement to mobilize the full passive resistance of the soil in the backfill.

On the front side of the wall, the initial soil thrust of 32.3 kN/m is not close to either minimum or maximum theoretical values of static soil thrust (active = 6.7 kN/m and passive = 95.5 kN/m, respectively). During the simulation, the soil thrust on the wall front stays within the range of 20 kN/m to 44 kN/m. Using the M-O method, the minimum and maximum theoretical values were calculated to be 8.7 kN/m and 92.0 kN/m, respectively. Neither the minimum nor maximum values of the soil thrust time history are close to their theoretical values, indicating that the deposit in front of the wall does not experience significant deformation to develop a full passive wedge when the wall is moving towards it, and the wall does not move away from it far enough to develop an active wedge at any time during the simulation. It can be noted

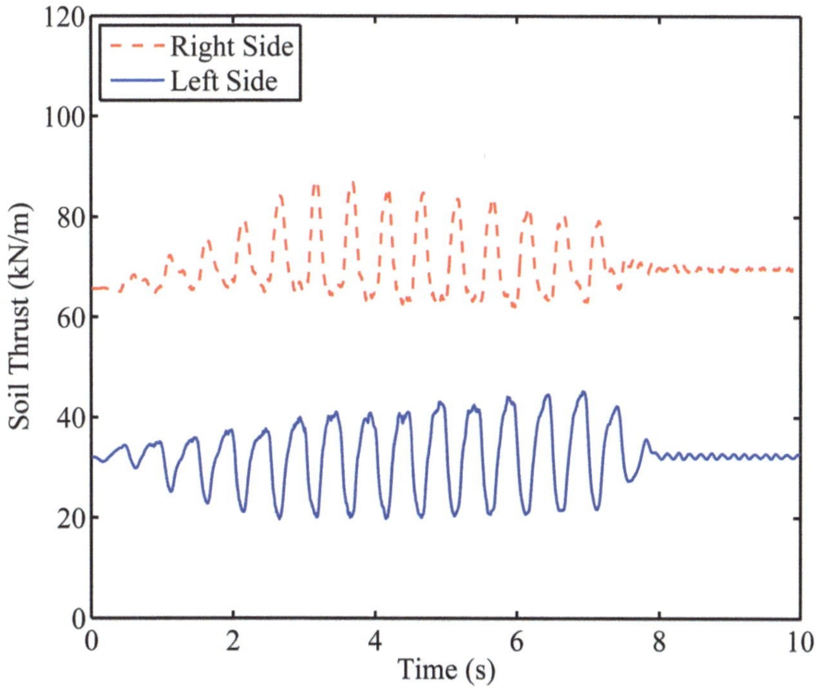

Fig. 3. Time histories of soil thrust on the retaining wall

that the soil thrust on the back of the wall (Fig. 3) has a 180o phase shift with the soil thrust on front of the wall indicating a side to side movement of the wall during the simulation. When the wall is moving away from the backfill, the soil thrust decreases on the back of the wall and increases on the front side of the wall.

Similarly to the soil thrust on the side of the retaining wall, forces acting on each particle in the base of the wall were tracked during the simulation allowing to determine the pressure distribution on the base of the wall. Special attention was paid to the pressure on the edges of the base wall, as rocking mode of motion would indicate development of higher pressures at the edges of the base during the dynamic loading. A portion equal to 10% of the length of the base wall was taken into consideration when monitoring pressure development on the tip and toe of the retaining wall (Fig. 4). Initially, the pressure under the toe is higher than the pressure under the heel due to initially higher thrust from the backfill of the wall. The edges of the base of the wall experience change in stress as time progresses. The 180o phase shift in the pressure was observed between the edges due to rocking. When the wall rotates away from the backfill the pressure on the tip increases and the pressure on the toe decreases, and when the wall rotates toward the backfill the pressure on the tip decreases when the pressure on the toe increases. Maximum value of the soil pressure developed at the tip of the wall is around 220 kPa, which is much smaller than predicted bearing capacity, therefore significant vertical displacement is not expected.

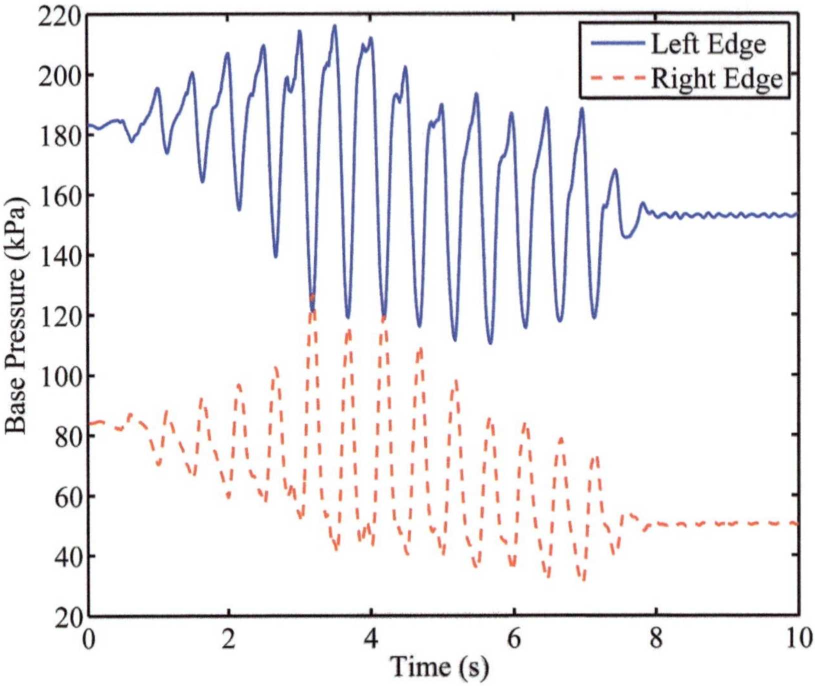

Fig. 4. Time-history of pressure on left and right edges of the base of the wall

Figure 5 shows horizontal, vertical and rotational displacement time histories of the wall. Horizontal displacement of the wall experiences cyclic as well as permanent modes of deformation as time progresses. These types of movement can be clearly seen in the displacement plot (Fig. 5a). It can be seen that displacement of the retaining wall consists of the accumulated displacement that increases as the simulation progresses and cyclic displacement that represents the side-to-side motion during every cycle. The total horizontal displacement the wall has experienced is 42 mm in prototype units, which is a little over the theoretical value of 40 mm using the method proposed by Richards and Elms (1979). The total settlement of the wall during the simulation was less than 1 cm in prototype units, and the total rotation of the retaining wall is less than 0.01 rad.

Changes in the magnitude and location of the soil thrust as well as changes in the distribution of the pressure on the base of the wall resulted in out of balance horizontal and vertical forces, as well as rocking moment at the center of the base of the wall. These forces were calculated throughout the simulation using the methodology presented by Zamani and El Shamy (2012). Figure 6 shows the moment-rotation relationships for this simulation. The moment-rotation history (Fig. 6a) shows high negative values of the moment were produced by the pressure from the backfill and caused the wall to rotate away from it. The accumulation of the rotational displacement during the simulation can be observed. The cyclic moment-cyclic rotation plot (Fig. 6b) shows that the moment-rotation relationship is nonlinear throughout the simulation, with rotational stiffness degrading as shaking progresses.

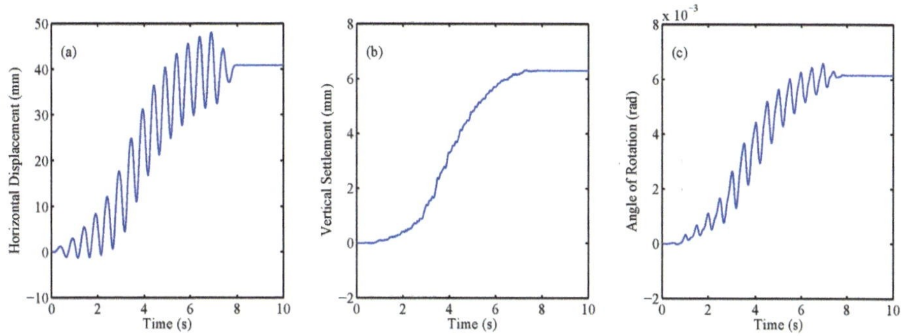

Fig. 5. Time histories of (a) horizontal displacement, (b) vertical settlement, and (c) wall rotation

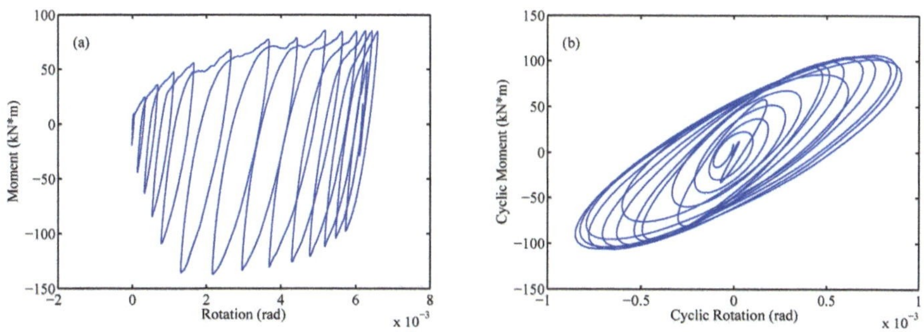

Fig. 6. Plots of (a) moment-rotation, and (b) cyclic moment-rotation

4 Conclusions

A DEM microscale approach is proposed to investigate the seismic response of the soil-retaining wall system in time-domain while taking into the account nonlinear behavior of the soil, possible separation between wall base and soil caused by rocking, sliding of the wall with respect to the ground, and dynamic soil-wall interaction. The results presented herein highlight the strength of the proposed DEM-based technique and its ability to model large-scale boundary value problems. The trends observed in the conducted simulations were similar to those of the existing experimental and analytical results. For instance, the simulation captured permanent wall displacement accumulation as well as rotational stiffness degradation as shaking progresses. Wall motion was amplified relative to the amplitude of input motion. Dynamic backfill thrust acting on the wall at the time instances when the wall is moving away from the backfill were comparable to the theoretical value of active earthquake pressure calculated using the M-O method. However, when the wall was moving towards the backfill it did not produce enough movement to mobilize the full passive resistance of the soil in the backfill. Results of additional simulations that examine the system response to various motion amplitudes and frequencies (including the response near resonance) are currently being compiled and will be published elsewhere.

Acknowledgments. This research was partially supported by the US Army Corps of Engineers Engineer Research and Development Center, grant number W9132 V-13-C-0004. This support is gratefully acknowledged.

References

Iai, S., Tobita, T., Nakahara, T.: Generalized scaling relations for dynamic centrifuge tests. Geotechnique **29**, 105–118 (2005)

Kramer, S., Smith, M.: Modified Newmark model for seismic displacements of compliant slopes. J. Geotech. Geoenviron. Eng. ASCE **123**(7), 635–644 (1997)

Mononobe, N., Matsuo, H.: On the determination of earth pressures during earthquakes. In: Proceedings of the World Engineering Congress, pp. 179–187, Oakland (1929)

Nadim, F., Whitman, R.: Seismically induced movement of retaining walls. J. Geotech. Eng. Div. **109**(7), 915–934 (1983)

Okabe, S.: General theory of earth pressures. J. Jpn. Soc. Civil Eng. **12**(1), 123 (1926)

Patsevich, A.: Discrete-element method analysis of seismic response of gravity retaining walls. MS thesis, Southern Methodist University, Dallas, TX, USA (2015)

Richards, R., Elms, D.: Seismic behavior of gravity retaining walls. J. Geotech. Eng. Div. ASCE **105**(GT4), 449–464 (1979)

Wartman, J., Bray, J., Seed, R.: Inclined plane studies of the Newmark sliding block procedure. J. Geotech. Geoenviron. Eng. ASCE **129**(8), 673–684 (2003)

Whitman, R., Liao, .S.: Seismic design of gravity retaining walls. Report No. Miscellaneous Paper GL-85-1, US Army Corps of Engineering (1985)

Zamani, N., El Shamy, U.: Analysis of the seismic response of soil-foundation-structure systems using a microscale framework. Soil Dyn. Earthq. Eng. **43**, 398–412 (2012)

Physical Modeling and Analysis of Site Liquefaction Subjected Biaxial Dynamic Excitations

Omar El-Shafee(✉), Tarek Abdoun, and Mourad Zeghal

Department of Civil and Environmental Engineering,
Rensselaer Polytechnic Institute, Troy, USA
{elshao2, abdout, zeghal}@rpi.edu

Abstract. The paper presents a series of centrifuge tests of level sites consisting of a granular soil deposits subjected to various bi-axial and uniaxial base excitations. The tests were conducted at RPI NEES centrifuge facility to assess the dynamic response characteristics of level deposits under multidimensional conditions. Synthetic sinusoidal waves were used as base excitations to test dense models under biaxial and uniaxial shaking. A dense array of accelerometers was used to monitor the deposit response along with pore water pressure transducers. One uniaxial test and one biaxial shaking test were conducted on two similar soil models to study the effect of multidirectional shaking on the generation of soil liquefaction for dense deposits. The observed acceleration and pore pressure are used along with non-parametric identification procedures to estimate the corresponding dynamic shear stress-strain histories. The measured results along with the obtained time histories are used to shed the light on the mechanisms of liquefaction occurring through the stratum under uniaxial shaking. It also shows the effect of soil relative density on soil behavior when subjected to biaxial shaking. This difference is evident in the coupling in stress-strain loops compared to the more known uniaxial soil behavior.

1 Introduction

The ability to understand and predict soil response during earthquakes is an important aspect in the design and management of soil-structure systems. Numerous studies were conducted over the last four decades to analyze and quantify the response of soil deposits subjected to base excitations. Most of the previous studies, especially those dealing with physical modeling were done using uniaxial base excitation, and only few used simple biaxial shaking.

Regarding experimental research, Ng et al. (2003) were the first to publish results of centrifuge test results using biaxial shaking. Their study was done on saturated granite embankments in Hong Kong, in which they found noticeable difference in settlement between biaxial and uniaxial tests. Su and Li (2006) used centrifuge testing (at Hong Kong university) to investigate the effect of multidirectional shaking on soil-pile interaction. They found that biaxial shaking produces higher relative displacement between pile and soil, and the pore water pressure is slightly higher. The final conclusion was that soil-pile interaction under multidirectional shaking cannot be separated

T. Abdoun and S. Elfass (eds.), *Soil Dynamics and Soil-Structure Interaction for Resilient Infrastructure*, Sustainable Civil Infrastructures, DOI 10.1007/978-3-319-63543-9_15

into two independent components due to coupling effects. Su and Li (2008) studied the effects of biaxial shaking on sand. They performed centrifuge tests using Toyoura sand in a laminar container and compared the results to constitutive model; the results showed that in biaxial testing pore water pressure rises faster and the soil has more settlement. Finally, Su (2012) tested piles under unidirectional and multidirectional lateral loading using biaxial motion platform in Shenzhen University, China and found that the pile resistance decreases under biaxial loading. The common theme in all these experimental studies was that the used biaxial shaking profiles were simple and mainly consisting of sinusoidal accelerations, in which the out of phase angle was kept constant throughout the experiment. Thus, the shake was not a real biaxial shaking changing direction randomly on the horizontal plane, but it was more like a uniaxial shake with a bias in a specific direction.

There were also a number of numerical simulations studies that used biaxial shaking. The available literature showed that biaxial shaking simulation was used mainly in the last three decades. One of the earliest attempts was the work done by Ghaboussi and Dikmen (1981), in which they developed a method to evaluate the seismic response and liquefaction by modeling soils as a two phase medium. Graves et al. (1998) investigated the effect of basin edge structure on amplifying ground response using biaxial ground motion simulations for site in Santa Monica (California). Bielak et al. (1999) worked on showing the importance of biaxial excitations. Skarlatoudis et al. (2012) used 3D finite difference model to study 3D wave propagation. Finally, Boaga et al. (2012) modeled biaxial wave dispersion in soil for some sites in Italy. Regarding the input motions used in numerical simulations, previous earthquake records were used in case of assessing damage for specific sites, or idealized earthquake records were used to match the requirements of design codes for the sites without any previous records. Another important aspect is how a uniaxial shaking can be used to represent the consequences of biaxial shaking. The arias intensity defined by Arias (1970) was used to ensure that the biaxial and uniaxial shakes used have the same energy content, which resulted in about 40% increase in the amplitude of the uniaxial input shake.

The work presented in this paper examined the validity of the assumptions used to simulate soil response during earthquake shaking, which are multidirectional in nature. This validation started by assessing the behavior and response of soil physical models subjected to biaxial base shaking. Followed by, uniaxial shaking test aimed at evaluating the validity of using equivalent uniaxial base shaking soil models to simulate biaxial shaking conditions. The recorded accelerations and pore water pressure were studied to evaluate the difference between the aforementioned testing scenarios. In order to better understand the differences in behavior the soil stresses and strains time histories were computed from the recorded accelerations inside the soil and compared directly.

2 Testing Procedure

The tested models of level site deposit consisted of a saturated Nevada 120 sand deposit 28 cm in height representing a prototype depth of 7 m at 25 g. The soil (Nevada 120) is clean uniform fine sand with minimum and maximum void ratios

Table 1. Centrifuge testing program

Test	D_r(%)	Model height (m)	Soil condition	Base shake
Test 2D	75	7	Saturated (viscous fluid)	2D
Test 1D	75	7	Saturated (viscous fluid)	1D

e_min = 0.55 and e_max = 0.751. The tests discussed in this article employed a deposit with relative density of 75%, representing a loose sand deposit Table 1.

The 2D laminar box used for studying the biaxial shaking effects on soils is made up of a stack of dodecagonal rings separated by roller bearings, arranged to permit relative movement between rings in two perpendicular directions plus rotation (X, Y and Θ) with minimal friction. A flexible strong latex membrane was used to contain the soil model to prevent leaking of soil and saturation fluid. The membrane was cylindrical in shape prepared out of thin latex sheet, and due to the size of the model the membrane was made by the researcher in the lab, a detailed procedure can be found in El-Shafee (2016).

The model was built using dry pluviation method, where the sand is placed in lifts, typically 2 cm thick, the soil layers were pluviated using a drop height (15−20 cm) with light tamping. The pluviation process was interrupted at pre-defined elevations for the placement of the accelerometers and pore water pressure transducers. The container was placed thereafter on the shaker platform (centrifuge basket). Saturation took place on the shaker platform to minimize disturbance to the soil deposit. Saturation was done by applying vacuum (28 inch mercury), then injecting CO2 to replace air in the voids. Vacuum pressure was reapplied followed by slowly saturating the model using viscous fluid (25 times viscosity of the water). Thus, the tested model corresponded to prototype permeability comparable to that of natural Nevada sand. Figure 1 shows the saturated model ready for spinning. When the g-level reached 25 g, the model was shaken with a series of biaxial or uniaxial base excitations.

The tested models represented a level site and were equipped with accelerometers, pore pressure transducers, laser displacement sensors and LVDTs. These sensors were mounted inside and outside the model at three elevations at depths of 1, 3 and 5 m measured from the model surface. Each elevation had the following sensors: Seven pairs of accelerometers to record the soil motion in the two horizontal X and Y directions, Two vertical accelerometers were used to monitor the vertical motion that maybe generated in the soil by the base excitation, Two pore pressure transducers were placed in each elevation to measure the excess pressures generated during the test. On the surface of the model there were three vertical LVDTs mounted to measure the surface settlement of the model.

The model was spun up to 25 g and then selected shaking events are applied to the base of the model using RPI 2D shaker. Two specific motions were used in this study shakes A and B. Both shakes were applied to the model in the following order; 0.01 g, 0.06 g, 0.01 g, 0.1 g, 0.01 g and 0.2 g amplitudes. This sequence was used for both shakes starting with shake A then shake B. After each shaking event the model was allowed enough time to dissipate any excess pore pressure before applying the following shake. The small nondestructive amplitude shakes of 0.01 g were used to compute the shear wave velocity profile for the soil model. The shear wave velocity measurements

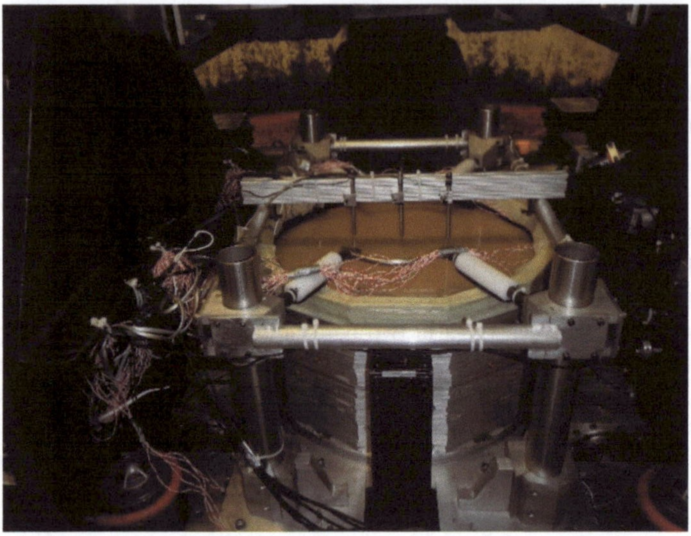

Fig. 1. View of the laminar box with fully instrumented model

were used to quantify the changes in soil properties after every shaking event. The large amplitude shakes were applied in ascending order to avoid disturbance of the model by applying the largest amplitudes in the beginning of the test. The equivalent energy uniaxial shakes with 40% increase in X component amplitude were applied in the same exact sequence. For the sake of simplicity the equivalent uniaxial shakes will be captioned with the biaxial shake amplitude (For example shake A 0.1 g in biaxial test is equivalent to 0.14 g in uniaxial test but the later will also be called "0.1 g"). The details of the shakes and methodology of developing them is presented in the following section. The name of the test will be followed by an abbreviation indicate which shaking profile is applied, for example biaxial test with profile A will be called "Test 2DSA".

3 Input Motions

Biaxial/Two Directional (2D) Shaking: The centrifuge soil model was subjected to a series of biaxial base excitations. The input consisted of synthetic motions with 1, 2, 3 Hz dominant frequencies and varying amplitude. The biaxial input motions were obtained by varying the phase angle θ of the y and x accelerations (θ = arc tan (a_y/a_x)), to give phase angle time history equivalent to real earthquake records. An example for earthquake record is provided in Fig. 2. It shows example of Lotung earthquake phase angle time history, while Figs. 3 and 4 shows the phase angle time history for shakes A and B respectively. Which shows although the used shakes were synthetic motions, the phase angle has some randomness and not far from the typical earthquake recorded phase angle time history. The variations of this angle showed a pattern consistent with that of real earthquake records. These two motions had the same frequency but with varying amplitudes mixed together.

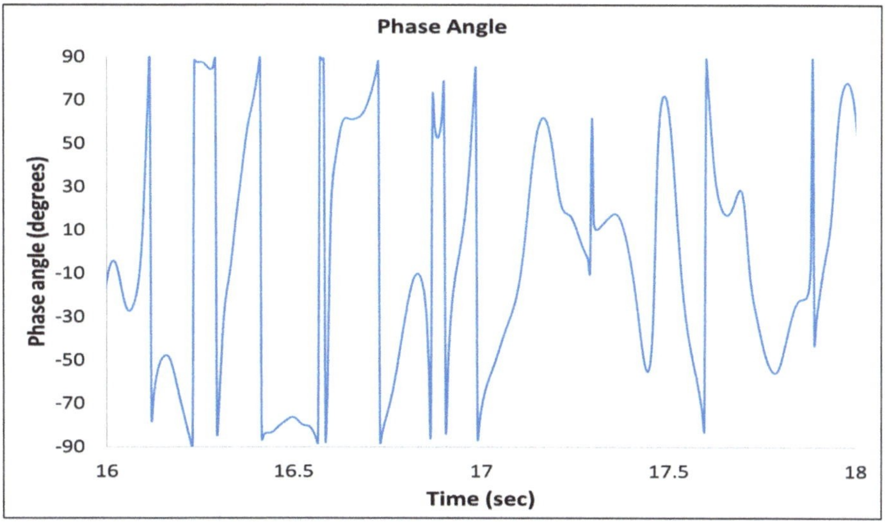

Fig. 2. Lotung earthquake record phase angle

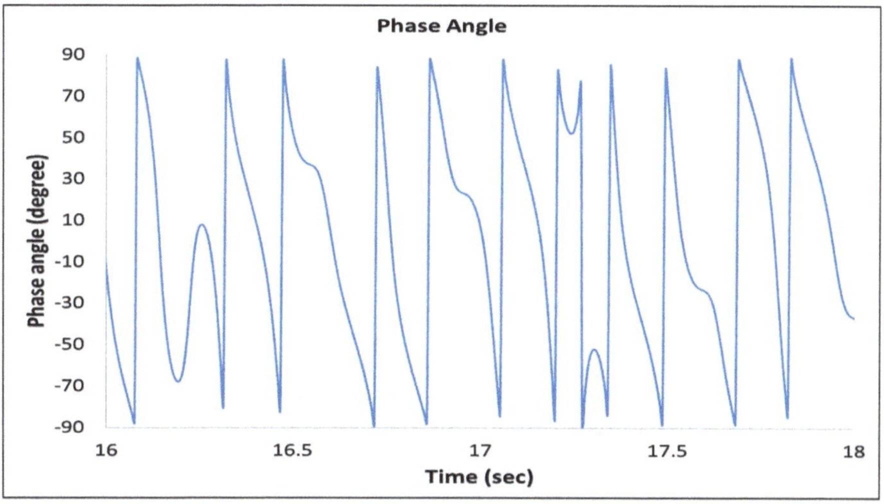

Fig. 3. Shake A biaxial phase angle

One Directional (1D) Shaking: The uniaxial shakes used in these tests have the same profile of the X component in the biaxial shakes. The uniaxial shaking profiles developed to have equivalent energy to the biaxial shaking. This was achieved using arias intensity so the uniaxial shake had the same input energy as the two components of the biaxial (2D) shake combined. The Arias intensity defined by Arias (1970) is the total energy per unit weight stored by a set of simple oscillators evenly spaced in frequency. The Arias intensity for ground motion in the x-direction (I_{xx}) can be written as:

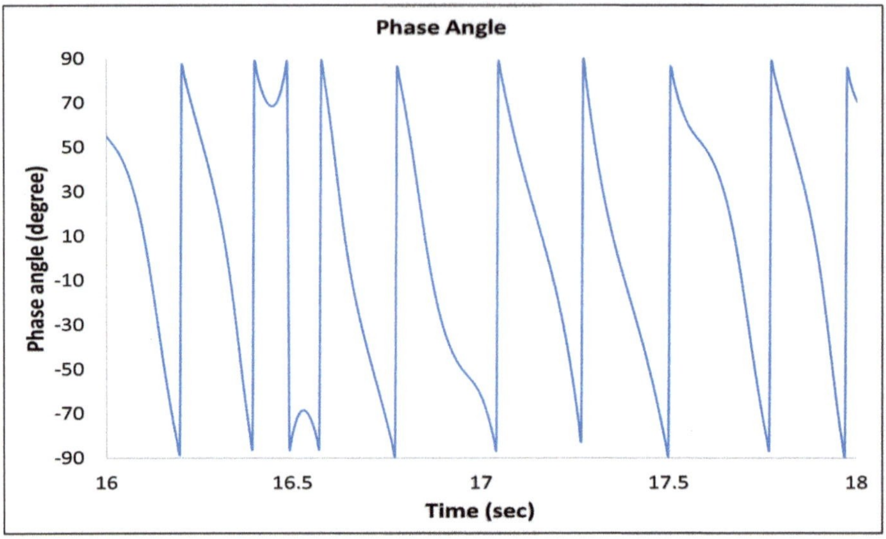

Fig. 4. Shake B biaxial phase angle

$$I_{xx} = \frac{\pi}{2g}\int_0^{T_d} a(t)^2 dt \qquad (1)$$

Where:

$a(t)$: The acceleration time history in the x-direction in (m/s^2).
g : The acceleration due to gravity in (m/s^2).
T_d : The total duration of motion in seconds.

The arias intensity in y-direction can be calculated similarly using the same quantities in Eq. 1 in the y-direction. To compare the biaxial base shake to the equivalent uniaxial base shake, the arias intensity of the uniaxial shake was equal to the resultant of the two components of the Arias intensity of the biaxial shake in x and y directions and can be written as:

$$I_{1D} = \frac{\pi}{2g}\int_0^{T_d} \left(a_x(t)^2 + a_y(t)^2\right) dt \qquad (2)$$

Where:

I_{1D} : Arias intensity for the equivalent uniaxial base shake.
$a_x(t)$: The acceleration time history in the x-direction in (m/s^2).
$a_y(t)$: The acceleration time history in the x-direction in (m/s^2).

Using Eqs. 1 and 2, the uniaxial equivalent energy shake was found to be 40% larger in amplitude. The uniaxial shake used was obtained by increasing the X component acceleration by about 40%. The reason for this 40% is that the two components of the biaxial shake are equal, thus it is the square root of two quantities of one (approximately equals 1.4), this was a special case because of the values of the selected biaxial motion. Due to a technical error the arias intensity of the biaxial shakes was about 20% smaller than the equivalent uniaxial shake.

4 Measured Soil Response

4.1 Nondestructive Shaking Accelerations

Nondestructive shakes (0.01 g) applied before and in-between large shaking events were used to calculate the shear wave velocity of the soil. Shear wave velocity is one of the methods used to verify that the tested models have the same initial soil properties. Another purpose for the shear wave velocity is to assess the change in soil properties in the two tests as the shaking sequence progress. Shear wave velocity was calculated using correlation technique Elgamal et al. (1995). This technique uses the acceleration records in two locations to calculate the time delay for the peak of the cross-correlation function, knowing the distance between these two locations the shear wave velocity is computed. The calculated initial shear wave velocity profile with depth for the three tests is presented in Fig. 5. It is clear that the velocities at depths of (Z = 2, 3 and 4 m) are almost the same for the two tests, which is a strong indication that the models built for the two tests are consistent and all have the comparable properties.

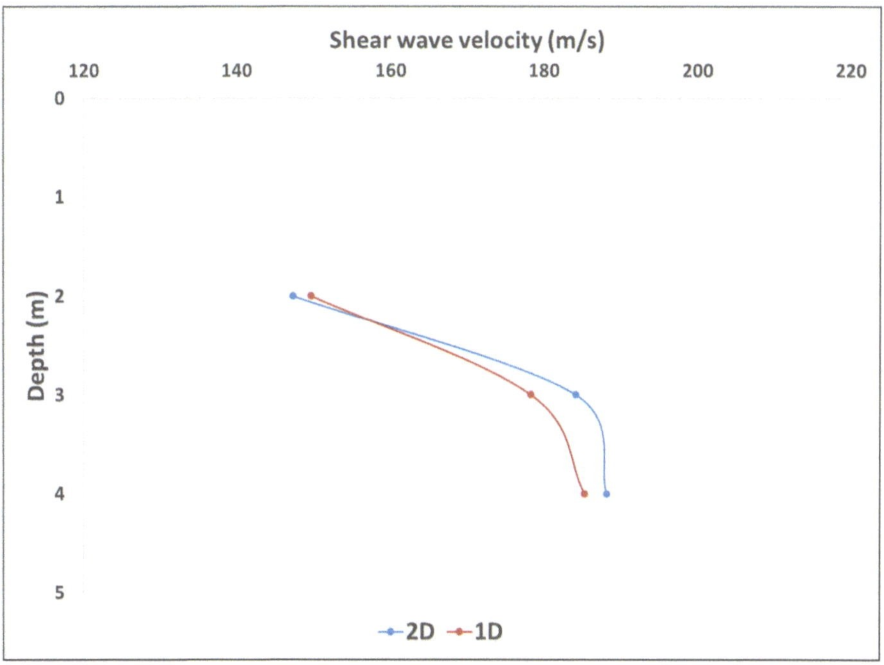

Fig. 5. Initial shear wave velocity profile for the two tests

4.2 Accelerations

Accelerations measured inside the soil under shake A with input amplitude 0.1 g are shown in Fig. 6 for the two tests. The results revealed that in Test 2DSA and Test 1DSA the amplitude of accelerations decreased up to 3 m and 1 m below the model surface respectively. This decrease in recorded acceleration amplitude is indicative of soil softening apportioning full liquefaction up to this depth. Closer look to the accelerations time history shows that in biaxial shaking the decay was faster than the uniaxial shake. This indicates faster liquefaction for the soil under biaxial shaking. Another observation that uniaxial test acceleration records shows large spikes up to 3 m below model surface. This reflects higher dilative behavior than that observed in biaxial shaking, which can be attributed to concentration of shear strain in one direction which will be explained in details in the subsequent section of stresses and strains. Similar trend was observed for larger amplitude of the same shaking profile (0.2 g), dilation spikes were observed in both tests especially at 1 m depth. The amplitude was slightly larger in Test 1DSA which was also accompanied by strain accumulation in uniaxial case. On the other hand, faster liquefaction in Test 2DSA is attributed to particle destabilization because of shearing in the perpendicular direction.

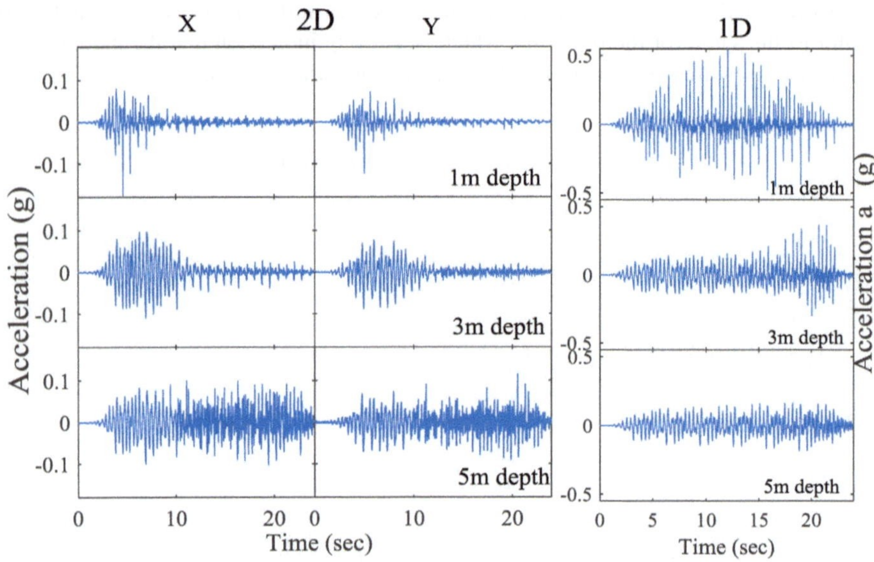

Fig. 6. Shake A 0.1 g recorded acceleration at different depths array 5

In order to confirm the trends and observations in the previous part of the section, results from shake B are presented here. The recorded accelerations using input amplitude 0.1 g are shown in Fig. 7. The results showed that for both Test 2DSB and Test 1DSB the acceleration decreasing trend was similar to shake A. Sand fully liquefied up to 1 m below the model surface. The decrease for both tests happened slightly faster in the biaxial test, which indicates that liquefaction happens almost in the same time in both tests. For Test 1DSB the same trend of large spikes occurring up to 3 m below the model surface was observed. But this time dilation spikes were monitored in Test 2DSB especially in X direction, however the amplitude of spikes is close to the ones measured in shake A. Similar to the previous shake, the trend continued for larger amplitude of shake B (0.2 g) caused large dilation spikes. This happened in both tests and it was very noticeable at 3 and 5 m depths with almost equal amplitude in both Test 2DSB and Test 1DSB.

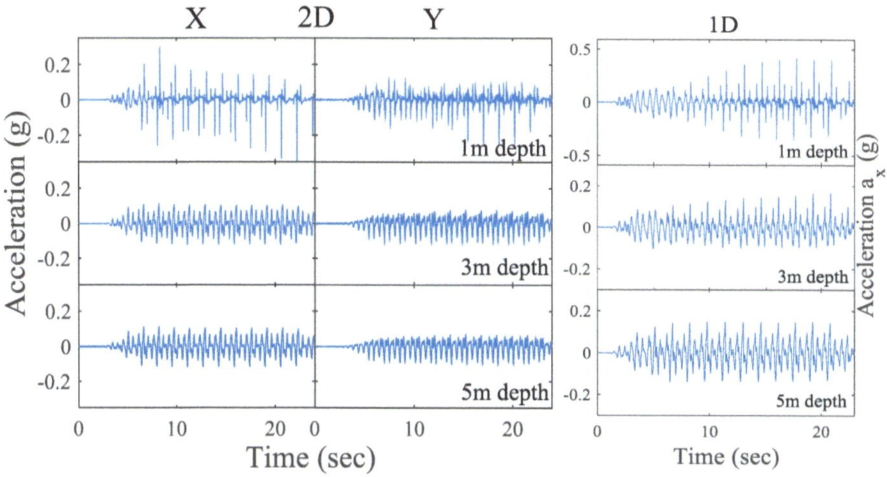

Fig. 7. Shake B 0.1 g recorded acceleration at different depths array 5

4.3 Pore Water Pressure

For excess pore water pressure the comparison is divided into two sections. First, comparison between Test 2DSA and Test 1DSA, then comparison between Test 2DSB and Test 1DSB. Regarding the first comparison Fig. 8 shows the pore pressure time history recorded during shake A with amplitude 0.1 g. With the exception of marginal differences in how fast the excess pore pressure rises and slightly larger spikes in Test 1DSA records, the two time histories are very comparable up to 3 m depth, the small differences are within the experimental error range. While for 5 m depth the pore pressure buildup is faster and larger for the biaxial shake. This indicated that although the soil properties was the same the response was different. It was also observed that for all depths uniaxial pore pressure records had large dilation spikes the slightly which is consistent with the larger spikes in the acceleration records. This consistency in differences between the two experiments increased the confidence in each individual soil response measurement. Similar trend was observed for shake B (0.1 g), except for larger spikes for uniaxial shake at shallow depth, and lower buildup for deep layers subjected to biaxial shaking. These differences did not come as a surprise given that the model was densified after the previous shakes and the fact that all the input energy is concentrated in one direction in the uniaxial test shake.

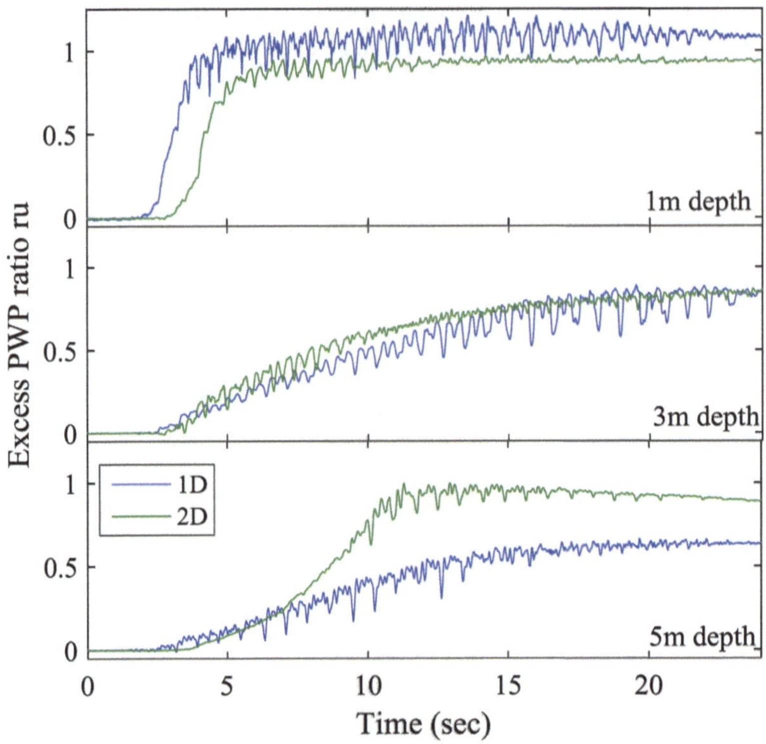

Fig. 8. Shake A 0.1 g PWP comparison

4.4 Calculated Stress-Strain Time Histories

The acceleration records for both tests were used to obtain the corresponding shear stress and strain vector generated inside the soil. The computed stresses were at the mid-points between the three sensors level. (i.e. 2, 3 and 4 m). These stresses and strains were evaluated using the system identification technique that was developed by Zeghal et al. (1993) and (1995). This technique provides non-parametric estimates of shear stress-strain histories utilizing only accelerations records provided by vertical arrays of accelerometers. For the tests discussed in this paper, the discussion of computed results will be mainly focused on the comparison of the loops for the two tests under shake B.

The stress strain loops for both test for shake B with 0.1 g amplitude are shown in Fig. 9. It was found that the biaxial test stress strain loops were comparable to the uniaxial test. The magnitudes of stresses were comparable for all depths and strains was almost the same at 2 m depth. The more important finding that for Test 2DSB the stress strain loops in X and Y directions were not independent and there was an obvious stress strain coupling. Figure 9 shows that for Y direction there was a big stress peak that was happening at almost zero strain. Since this observation needed a closer look the loops discussed here was broken into pieces and each segment was one cycle. These segments were studied to see what happens in each direction at the same time.

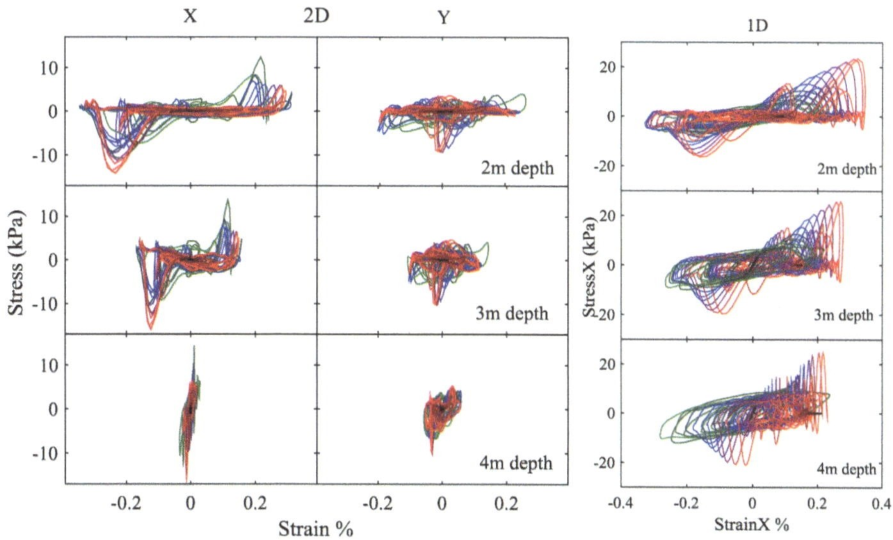

Fig. 9. Dense model shake B 0.1 g 2D and 1D stress strain loops (coupling loops)

Figure 10 is showing sample of these cycles plotted with color mapping to help locate the instance in which coupling happened. It is obvious that at 2 and 3 m depth the stress peaks in X direction happened at strains of 0.1% and 0.2% while the corresponding stress peaks in Y direction happened at almost zero strain. This coupling resulted from the non-proportional loading conditions happening in biaxial shaking test and needs more and deeper analysis to be fully understood, more details can be found in the work done by El-Shafee et al. (2016).

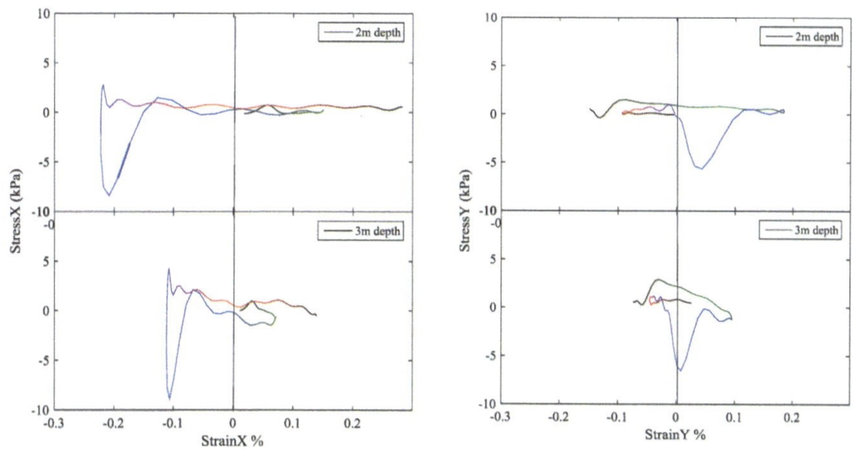

Fig. 10. Shake B 0.1 g 2D X-Y stress-strain coupling

5 Conclusion

This paper presents the measured results of centrifuge tests conducted to assess the effects of a biaxial shaking on the response of a level ground deposits, and comparing it to uniaxial excitations. The recorded accelerations were used to evaluate the two components of the shear stress and strain vectors acting on the horizontal planes of the tested deposits. Along with comparisons of shear wave velocity and excess pore water pressure measured in the soil. The measured relative density and computed shear wave velocity strongly indicated that the two tested soil models had similar soil condition at the beginning of the test. The change in shear wave velocity after the first shake was small for the two tests.

The recorded soil response indicated that there are differences between the two tests discussed in this paper. The main difference in recorded soil response was how fast the acceleration amplitude decrease inside the soil, and the presence of dilation spikes in both acceleration and pore pressure records. The higher dilative behavior in Test 1D (especially in the later runs) was mainly due to input energy concentration in one direction, versus distributing it on a plane in the case of Test 2D. Also the pre-shaking of the soil has an impact on the results, earlier shakes had larger liquefaction depth for Test 2D, while the opposite was observed after subjecting the model to several shakes and densifying it.

The computed stress-strain loops showed that for later shakes the magnitudes of stresses were comparable for all depths and magnitudes of strains were comparable at shallow depth. These latter stresses and strain showed a complex response pattern associated with coupling of the soil response in the two horizontal X and Y directions during phases of large strains and associated dilative response.

The results of the two tests leading to the following conclusion. Although the two tests had the same energy and conducted on the same soil conditions. The response for dense deposits was not the same with higher level of complexity in the biaxial case especially when analyzing the stresses and strains in the soil due to the coupling between the two directions. It is recommended to use the biaxial shaking, because the equivalent uniaxial is lacking the ability to capture the multiaxial response of the system which can be vital in complex problems.

References

Arias, A.: A measure of earthquake intensity. In: Hansen, R.J. (ed.) Seismic Design for Nuclear Power Plants, pp. 438–483. MIT Press, Cambridge (1970)

Bielak, J., Xu, J., Ghattas, O.: Earthquake ground motion and structural response in alluvial valleys. J. Geotech. Geoenvironmental Eng. **125**, 413–423 (1999)

Boaga, J., Renzi, S., Vignoli, G., Deiana, R., Cassiani, G.: From surface wave inversion to seismic site response prediction: beyond the 1D approach. J. Soil Dyn. Earthq. Eng. **36**, 38–51 (2012)

Elgamal, A.W., Zeghal, M., Tang, H.T., Stepp, J.C.: Evaluation of low-strain site characteristics using the Lotung seismic array. J. Geotech. Eng. **121**(4), 350–362 (1995)

El-Shafee, O.: Physical and computational modeling of biaxial base excitation of sand deposits. Doctoral dissertation, Available from ProQuest Dissertations and Theses database (2016)

El-Shafee, O., Zeghal, M., Abdoun, T.: Analysis of site liquefaction subjected to biaxial dynamic base excitation. J. Soil Dyn. Earthq. Eng. (2016). (to be submitted)

Ghaboussi, J., Dikmen, S.U.: Liquefaction analysis for multidirectional shaking. J. Geotech. Eng. Div. ASCE **107**(GT5), 605–627 (1981)

Graves, R.W., Pitarka, A., Somerville, P.G.: Ground-motion amplification in the santa monica area: effects of shallow basin-edge structure. Bull. Seismol. Soc. Am. **88**(5), 1224–1242 (1998)

Ng, C.W.W., Li, X.S., Van Laak, P.A., Hou, D.Y.J.: Centrifuge modeling of loose fill embankment subjected to uni-axial and bi-axial earthquakes. J. Soil Dyn. Earthq. Eng. **24**, 305–318 (2003)

Skarlatoudis, A.A., Papazachos, C.B., Theodoulidis, N.: Site-response study of Thessaloniki (Northern Greece) for the 4 July 1978 M 5.1 aftershock of the June 1978 M 6.5 sequence using a 3D finite-difference approach. Bull. Seismol. Soc. Am. **102**(2), 722–737 (2012)

Su, D., Li, X.S.: Centrifuge modeling of pile foundation under multi-directional earthquake loading. In: Physical Modelling in Geotechnics-6th ICPMG, vol. 1, pp. 1049–1055 (2006)

Su, D., Li, X.S.: Impact of multidirectional shaking on liquefaction potential of level sand deposits. J. Geotech. **58**(4), 259–267 (2008)

Su, D.: Resistance of short, stiff piles to multidirectional lateral loadings. J. Geotech. Test. **35**(2), 1–17 (2012)

Zeghal, M., Elgamal, A.W.: Lotung site: downhole seismic data analyses. Report, Civil and Environmental Engineering Department, Rensselaer Polytechnic Institute, Troy, NY (1993)

Zeghal, M., Elgamal, A.W., Tang, H.T., Stepp, J.C.: Lotung downhole array. II: evaluation of soil nonlinear properties. J. Geotech. Eng. **121**(4), 363–378 (1995)

Assessment of Earthquake Induced Lateral Displacements at Transpower Hayward HVDC Link Pole 3Upgrade

Ian McPherson(✉)

Ground Engineering Aurecon New Zealand Limited, Wellington, New Zealand
ian.mcpherson@aurecongroup.com

Abstract. Transpower New Zealand Ltd owns and operates the High Voltage Direct Current (HVDC) link which transfers power between the North and South Islands of New Zealand. The link is a critical lifeline in the electrical distribution system. Transpower proposed upgrading the Haywards facility at the north end of the HVDC link from 700 MW to 1200 MW. The development included construction of a new convertor station (the Pole 3 building) which was located within 150 m of the active Wellington Fault. This fault could generate a Magnitude 7.3 earthquake with a 1 in 2,500 year return period peak ground acceleration of 1.01 g. The terrain in the area of the substation is relatively steep and the site was developed by cutting into higher ground and placing the material as un-engineered fill into the gullies. The new Pole 3 building straddles the cut fill interface with rock at shallow depth at one end and up to 25 m of uncontrolled fill at the other end. A key geotechnical design issue was to assess how far the underlying fill could move in the design earthquakes so that the base isolators in the new building could be designed. This paper outlines the analytical methods to assess likely lateral fill movements.

1 Introduction

Transpower NZ Ltd owns and operates the High Voltage Direct Current (HVDC) link which connects the electrical systems in the North and South Island of New Zealand. Transpower proposed upgrading the link from 700 MW to 1200 MW with a new HVDC converter station (the Pole 3 Building) at the Haywards (North Island) end of the link. The facility is a key element of the electricity transmission network in New Zealand and it must remain operational under large earthquakes and be repairable even after a 1 in 2,500 year event.

Transpower engaged Siemens NZ Ltd to carry out the upgrade with engineering services, including geotechnical engineering, provided by Aurecon New Zealand Ltd. A key part of site development from a geotechnical perspective was assessing likely lateral movement during very large earthquake and designing a building and foundation system to manage large lateral displacements. The new foundation system had to accommodate a base isolation system designed to minimise the effect of strong ground shaking on the building and its contents.

T. Abdoun and S. Elfass (eds.), *Soil Dynamics and Soil-Structure Interaction for Resilient Infrastructure*, Sustainable Civil Infrastructures, DOI 10.1007/978-3-319-63543-9_16

This paper discusses geotechnical design of the new Pole 3 building and methodology used to assess the potential movement of un-engineered fill during large earthquakes. It describes site conditions, summarises the site geotechnical information, and presents the methodology used to derive a site geotechnical model, The paper then presents the results of stability analyses using a range of analytical methods including two dimensional static and pseudo static, three dimensional static and pseudo static, Newmark sliding block analyses, and finite element modelling. It also discusses the results of the analyses and draws conclusions on the likely magnitude of horizontal ground displacements which formed the basis of building and foundation design.

2 Site Description

The site has been developed since the 1950s with extensive terrain re-shaping by cutting the higher ground and infilling gullies to create level building platforms. Earthworks included infilling a large, deeply incised, 'Y' shaped gully system (See Fig. 1). The maximum fill depth in the main gully is up to 35 m deep (See Figs. 2 and 3 for original and current contours).

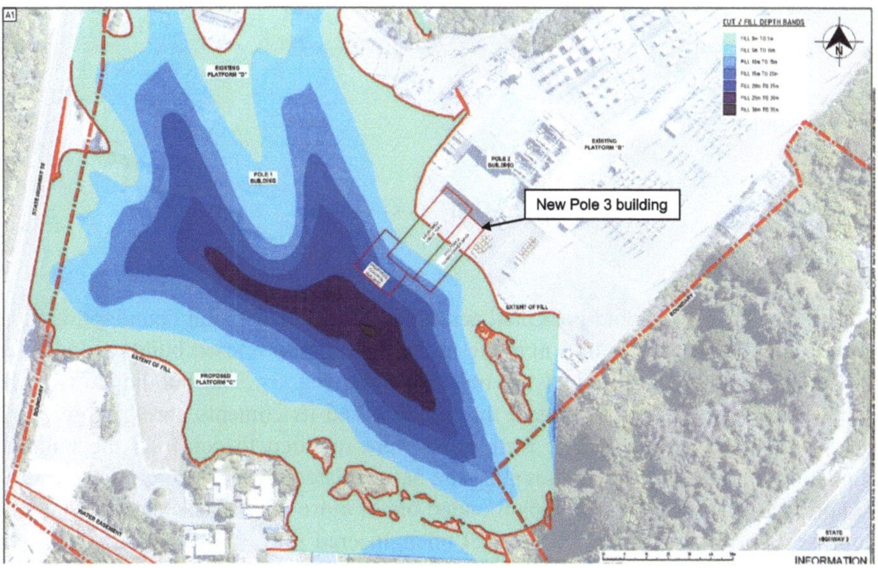

Fig. 1. Gully depth of fill contours

The site is underlain by greywacke rock which is variably weathered and is pervasively jointed due to tectonic movements. The fill derived from the greywacke ranged from essentially unweathered, moderately strong, angular and sub angular gravels and cobbles through to silty soils derived from completely weathered rock. Because the fill was excavated from different parts of the site, and was laid down in layers, it is highly variable both vertically and horizontally.

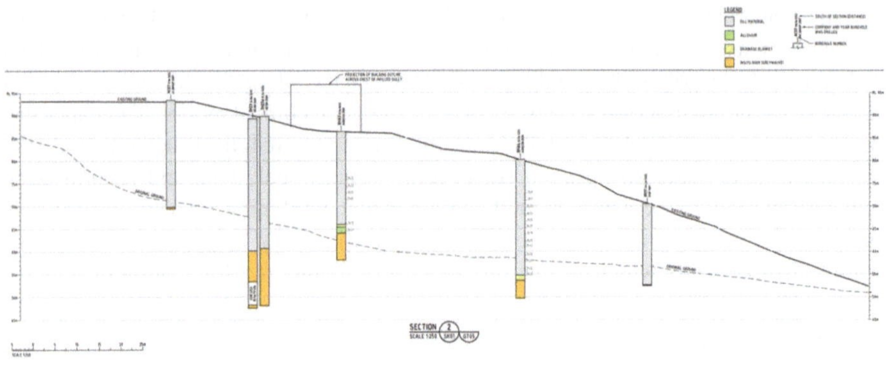

Fig. 2. Indicative long section down gully

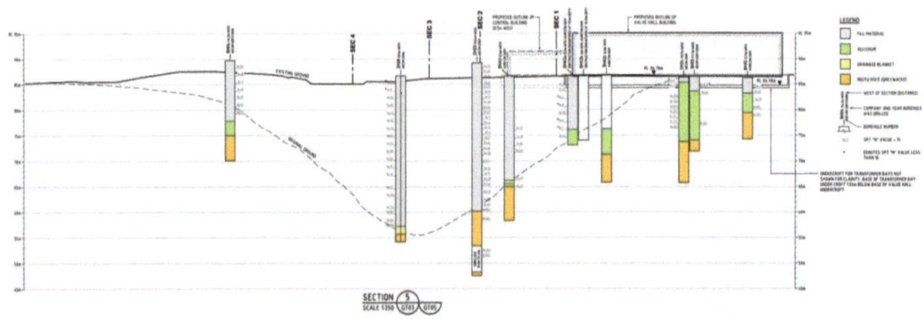

Fig. 3. Gully cross-section

The new Pole 3 building is located across the fill/natural ground boundary with the depth to rock between 3 m to 4 m at the north end of the proposed building and 25 m to 30 m under the south end of the building. (See Fig. 3). The building was to be founded on base isolaters to protect the building and its contents during large earthquakes, and excavations 3 m to 6 m depth were required to construct the building foundations and undercroft for mounting the base isolaters. After undercroft excavation the north east part of the building is founded on, or close to, the rock surface while the south west end is founded on 20 + m of unengineered fill.

The potential for the fill to displace under earthquake loading was recognised as a key hazard and preliminary calculations indicated that several hundred millimetres of movement could occur due to movement on the adjacent segment of the Wellington Fault. The Wellington Fault is located approximately 150 m east of the site and a seismic hazard assessment indicated that a peak ground acceleration of 1.01 g could occur during a 1 in 2500 year event.

3 Review of Available Information

Geotechnical investigations have been carried out across the site over the last 30 to 40 years including, 38 boreholes and laboratory testing in 2009, 10 boreholes in 2008 and 23 boreholes and 19 cone penetrometer tests in 1987.

Despite the previous investigations covering a large portion of the site there was limited information on the depth to rock in the area of the proposed Pole 3 building. To provide additional information in key areas Aurecon undertook additional investigations including five 31 m deep boreholes with measurement of the undrained shear strength of the recovered core using a pocket penetrometer. Downhole shear wave velocity testing was carried out in two boreholes and laboratory testing was carried out on selected core samples of gully fill. The testing was biased to cohesive soils as gravelly soils could not be tested by either shear box or triaxial testing. Four triaxial tests were carried out to provide effective stress strength parameters and 17 quick shear box tests to provide additional information on the undrained shear strength of the gully fill.

The triaxial tests were carried out in stages at three consolidation pressures with an initial consolidation pressure approximately equal to 50%, 100% and 200% of the sample's insitu vertical effective stress. The shear box test results are plotted in Fig. 4a. The confining pressure used during shear box testing was approximately equal to the sample's in-situ vertical effective stress calculated from the sample depth and assuming a soil unit weight of 20 kN/m^3.

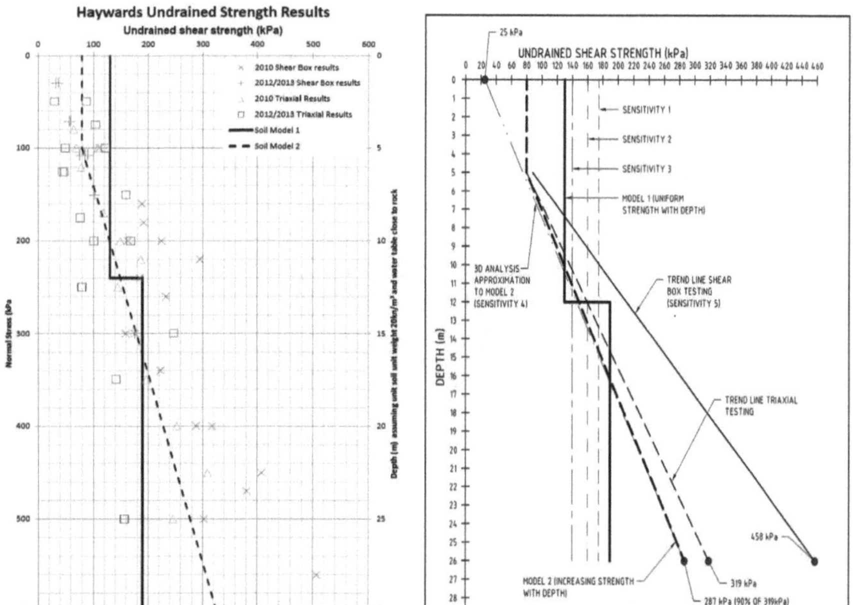

a)Shear Box and Triaxial Test Results **b) Developed Soil Strength Models**

Fig. 4. Soil strength models

Samples were tested in the shear box at their in-situ moisture content rather than being saturated. This is because the soils are partially saturated in situ and partially saturated soils will generally be stronger than saturated soils.

The pocket penetrometer readings enabled a substantial quantity of data to be captured (141 readings). The undrained shear strengths from pocket penetrometer testing were constrained at 300 kPa as this was the maximum value that could be measured with this instrument. The strength of approximately 25% of the samples tested exceeded the capacity of the instrument so the mean undrained shear strength values derived from pocket penetrometer testing may be lower than the overall mean strength of the soils underlying the site.

4 Geotechnical Model

4.1 Soil Strength Parameters

Undrained shear strength parameters were assumed to be applicable for earthquake analysis as earthquake impose rapid loads with limited time for dissipation of excess pore pressures. The soil undrained shear strength was derived based on four sources of information: pocket penetrometer testing on borehole samples, triaxial testing, shear box testing and Standard Penetration Testing (SPT) results.

The pocket penetrometer results showed limited change with depth whereas triaxial and shear box testing show a strong increase with depth. Usually soil undrained shear strength increases with depth and increasing vertical stress but possibly the compaction applied during fill placement effectively over-consolidated the soil. Because of the uncertainty in likely shear strength with depth profile the four sources of information were used to derive two soil strength profiles as follows:

- A two-layer model based on the shear box test results with a constant 130 kPa undrained shear strength down to 12 m depth and 190 kPa below 12 m depth.
- A bilinear model with a 5 m thick upper layer with an undrained shear strength of 80 kPa and undrained shear increasing with depth based on the triaxial test results. The design strength profile was factored down by 10% to allow for any cyclic softening during earthquake related ground shaking

Plots of the two soil profiles are presented in Fig. 4b. A range of sensitivity analyses with respect to soil strength were carried out using four soil strength profiles as shown in Fig. 4b.

For the underlying rock, the following parameters were adopted

- Cohesion 20 kPa
- Friction angle 40°
- Bulk Density 24 kN/m^3

The cohesion and friction values were selected as reasonable values which would form a strength boundary with the overlying fill soil to ensure that failure slip plane occurs within the overlying soil fill. The bulk density was based on a typical value for Wellington Greywacke rock.

4.2 Groundwater

Measured water levels in boreholes located towards the top of the infilled gully indicated water at 13 m to 20 m depth or close to rock level. A 1987 investigation report (Beetham and Fellows 1987) stated that water levels would be close to rock which was assessed as reasonable as the steep topography would shed water relatively quickly. In addition, the results from a borehole located more or less on the axis of the old gully had 2 m to 3 m of gravel, inferred to be a drainage blanket placed as part of gully infilling, on top of the rock. This gravel was highly permeable and was difficult to seal off with grout at the time of drilling and when the piezometer was being installed. Such a drainage blanket, if it extended under the entre gully fill, would draw water levels down and hence a water level following the top of the rock surface was adopted for design.

4.3 Seismic Loads

The Transpower brief specified earthquake peak ground accelerations (PGA) for three return periods (1 in 500 years, 1 in 1,000 years and 1 in 2,500 years) and for two types of site, underlain by rock and underlain by a shallow thickness of soil. The PGAs for a 1 in 2,500 year event were 0.75 g and 1.01 g for rock and soil foundations respectively.

Based on the equation that site period = $4H/V_s$ (where H = depth to rock and V_s = shear wave velocity), the site period varies from 0.1 s to 0.15 s at the north end of the building and from 0.20 s to 0.25 s at the south end of building. Both site periods are less than the New Zealand loadings code definition of a Class C Shallow Soil site which is 0.6 s. Hence, the use of a 1.01 g PGA (i.e. assuming that soil underlies the building with rock at depth) may be overly conservative. The theoretical increase in site period as the depth of fill increases from north to south was one of many factors which had to be considered when assessing the magnitude of lateral movements under large design earthquakes.

4.4 Earthquake Selection

The Newmark sliding block model and finite element modelling required the use of real earthquake accelerograms. The choice of earthquakes for use in the analysis was based on three criteria:

- a large PGA to minimise scaling of the accelerogram to the design level acceleration.
- a rock site as rock accelerations were required for inputting into the base of the model for the Newmark sliding block and finite element analyses.
- an earthquake with similar attributes to the Wellington Fault earthquakes (i.e. site less than 1 km from the fault, forward directivity and slip strike).

Two large earthquakes were adopted (Lucerne 1992 and Tabas 1978) to minimise the need for large scale scaling of PGA. Properties of the two design earthquakes are summarised in Table 1 below.

Table 1. Summary of earthquake accelerograms used in analysis

Earthquake	PGA	Magnitude	Distance	Site description	Attributes
Lucerne 1992	0.82	7.3	1.1 km	Soil and alluvium	Strike-slip with forward directivity
Tabas 1978	0.93	7.4	1.2 km	Sedimentary rock	Strike-slip with forward directivity

4.5 Geometry of the Analytical Model

The geometry of the infilled gully is three dimensional with the depth of fill varying both across the gully and in the up and down directions. The depth of fill contours are shown in Fig. 1 and the original and current ground contours are presented in Figs. 2 and 3. Four sections were selected for detailed analyses.

5 Geotechnical Analysis and Results

5.1 General

The site is very complex with a three dimensional gully containing fills of variable grading and strength and hence four methods were used to provide an indication of likely lateral movements under seismic loading as follows:

- A pseudo static analysis was carried out to assess the yield acceleration (PGA to give a factor of safety of one) and to identify critical surfaces for each of the four design cross-sections. Displacements were then calculated by comparing the yield acceleration with the design PGAs using the methods of Ambraseys and Menu (1988), Bray and Travasarou (2007).
- Displacements were calculated using the Newmark Sliding method in conjunction with scaling the two design earthquakes to the design PGAs. Different earthquakes gave different displacements which gave an indication of variability due to varying accelerograms.
- Finite element modelling was used to calculate lateral displacements using the two design earthquake PGAs.
- A three dimensional stability analysis was used to assess the beneficial effects of shape of the gully.

The 2D analyses were carried out using software from Geo-Slope International with Quake/W used to calculate initial stresses in the ground and dynamic stresses from the earthquake, Sigma/W used to calculate seismic strains and displacements using the stresses from Quake/W, and Slope/W used to calculate static and pseudo-static stability plus displacements using the Newmark sliding block method in conjunction with Quake/W output.

5.2 Results

The results of the various analyses are summarised below.

5.2.1 Two Dimensional Pseudo Static

Undrained two dimensional pseudo static analyses were carried out to provide an indication of the stability of the gully fill under the specified PGAs and to assess likely ground movements at the four design cross-sections. The ground movements were assessed using the methods of Ambraseys and Menu (1988), Bray and Travasarou (2007) as discussed below.

Ambraseys and Menu (*1988*)

The method of Ambraseys and Menu uses the ratio of the yield acceleration (i.e. the acceleration which gives a Factor of Safety of 1) and the design PGAs to calculate indicative ground movements. The results indicated that the likely displacements increase with increasing depth of filling with the greatest movements likely along the centre of the infilled gully, reducing up each side of the gully as the depth of filling decreases. Minimal ground movement is likely beneath the northern portion of the valve hall and transformer bays with small ground displacements beneath the south end of the control building. The building is expected to rotate slightly in the plane of its floor slab due to the differential movement from north to south.

Displacements under a 500 year event are assessed to be less than 10 mm under the building and possibly up to 60 mm in the centre of the gully, and as unlikely to exceed 70 mm displacements during the 1000 year event, where service must be restored within five days. Assessed ground displacements increase up to 210 mm under a 2500 year event.

Bray and Travasarou (*2007*)

The method of Bray and Travasarou (2007) assesses displacements using the earthquake magnitude, the yield acceleration, the fundamental site period and the spectral acceleration at 1.5 times the fundamental site period.

The fundamental site period is difficult to assess for the Haywards site as it is a function of soil depth and shear wave velocity. The depth of moving ground in the zone of interest ranged from about 13 m to 28 m and the shear wave velocity ranged from about 250 m/s at the ground surface up to about 400 m/s below 10 m depth. Hence, the spectral acceleration for the site varied between 1.5 g and 2.0 g, for deeper surfaces just above rock level while movement along shallower surfaces could be driven by spectral accelerations between 2.0 g to 2.9 g.

Considering all the factors above, and using yield accelerations from the three dimensional modelling, it was assessed that lateral displacements for deep surfaces close to rock and extending upslope of the new building could vary between 30 mm to 350 mm. The lower value is based on rock spectral accelerations and is the mean less one standard deviation movement. The upper value is based on soil spectral accelerations and is the mean plus one standard deviation.

5.2.2 Three Dimensional Analyses

The gully fill is a three dimensional body of soil as shown in Fig. 1 with a "Y" shape and 0 m to 35 m of fill. Three dimensional effects increase stability with for example, soil from the two arms of the side gullies "jamming" together rather than moving smoothly down the axis of main gully. There would also be friction and cohesion along the sides of the gully fill if it moved down slope which are not allowed for in a two dimensional analysis.

To provide an indication of the increase in stability when allowing for three dimensional effects (or under assessing fill stability when using two dimensional analyses), a series of three dimensional analyses carried were out using the software package TSlope. The shape of the three dimensional failure surface was based on the geometry of the two dimensional analyses and of the valley. The software requires that the three dimensional failure surfaces are fully defined so the bulk of the failure surface was modelled as following the inferred rock level with a 5 m deep tension crack adopted to limit the north and south extents of the failing soil. The critical failure surface was identified by modelling failure in the lower part of the gully and incrementally moving the head of the potential failure zone upslope until a minimum value for the yield acceleration was found.

The TSlope and Slope/W results were calibrated by carrying out a 2D analyses along one section with the factors of safety from the two methods within 1%. Hence it was considered that the two dimensional results from Slope/W can be realistically compared with three dimensional stability results from TSlope.

For the seismic analysis, a number of analyses were carried out with the head of the failure surface moved from downslope to upslope of the new building in 10 m intervals. The yield acceleration from the inferred critical surface is presented in Table 2 below and is significantly higher than two dimensional yield accelerations (i.e. 0.61 g cf 0.42 g).

5.2.3 Newmark Sliding Block Analysis

The Newmark sliding block analysis is a commonly used method to assess displacement under earthquake loading. The method compares the yield acceleration of the sliding block to the acceleration in the earthquake accelerogram with the block of soil moving whenever the ground acceleration exceeds the yield acceleration. The benefit of this method is that it uses real earthquake data, which has a number of cycles of varying acceleration over time, whereas the pseudostatic model only uses the peak ground acceleration.

The process comprised using Quake/W to determine initial stresses in the ground and then the dynamic stresses in the ground caused by the earthquake specific accelerogram. The next step was to run the Newmark sliding block analysis in Slope/W which identified a number of critical failure circles and calculated factors of safety and displacement at 0.02 s intervals using the dynamic stresses from Quake/W. The analysis results are presented in Table 2 below. The displacements calculated are largest for the Tabas earthquakes as this had the largest number of higher acceleration peaks.

Table 2. Summary of Displacement Analysis Results for Sect. 2

Method	Earthquake	Yield acceleration	Design acceleration					
			0.54 g (500 year event)		0.74 g (1000 year event)		1.01 g (2500 year event)	
			95%[(a)]	50%[(b)]	95%[(a)]	50%[(b)]	95%[(a)]	50%[(b)]
Model 1 "Uniform" Strength with Depth								
Pseudostatic		0.42 g	10 mm	0 mm	70 mm	20 mm	210 mm	50 mm
Ambraseys and Menu (1988) Bray and Travasarou (2007)							30 mm to 350 mm	
Newmark sliding block	Lucerne 1992			0 mm		0 mm		16 mm
	Tabas 1978			12 mm		46 mm		69 mm
Finite element	Lucerne 1992			1 mm		5 mm		8 mm
	Tabas 1978			4 mm[(d)]		−6 mm		16 mm
Three dimensional		0.61 g	0 mm	0 mm	0 mm	0 mm	50 mm	10 mm
Model 2 Increasing Strength with Depth								
Pseudostatic		0.45 g	10 mm	0 mm	50 mm	10 mm	180 mm	40 mm
Newmark sliding block/Finite element	No results as software requires uniform strength within a soil layer							
Three dimensional		0.52	0 mm	0 mm	20 mm	10 mm	110 mm	30 mm

Notes: [(a)]Probability of movement exceeding predicted value is 5% using pseudostatic method. [(b)]Probability of movement exceeding predicted value is 50% using pseudostatic method. [(c)]All movements rounded to nearest 10 mm and taken as zero if yield acceleration is within 10% of design acceleration. [(d)]Negative movement uphill.

5.2.4 Finite Element Modelling

For finite element modelling the analytical cross-sections used in the Slope/W analysis were exported to Quake/W to ensure that consistent models were used in pseudo static and finite element modelling with respect to site stratigraphy and soil strengths. The Lucerne (1992), and Tabas (1978) earthquakes were used as they had large peak ground accelerations which minimised the scaling required to match the design level acceleration with scaling factors for the 1 in 2,500 year event of 0.81, and 0.91 for the Lucerne and Tabas cases respectively. The records had two orthogonal horizontal records and the record with the highest peak ground acceleration was used.

The static initial and dynamic stresses were calculated using the Quake/W software package and these stresses are then exported into Sigma/W to calculate the dynamic deformations for the model. A range of fixed and free boundary conditions were used to assess the effect of varying boundary conditions.

An equivalent elastic model for the fill material was used as there was a lack of any accurate stress-strain deformation information to justify undertaking a more complex non-linear dynamic analysis. The Model 1 soil strength profile only was used as Quake/W does not have an algorithm for undrained soils that vary in strength with depth. Elastic moduli values were assessed based on shear wave velocity test results with Poisson's ratio values, the fill type, and a review of published values appropriate for the silty sandy gravelly soils underlying the site.

The calculated residual lateral displacements are summarised in Table 2 below. The analytical results indicated that the fill could move upslope and downslope depending on the magnitude and direction of each earthquake pulse with generally a small residual downslope displacement at the end of strong ground shaking. A maximum displacement of approximately ±90 mm (i.e. upslope and downslope) was indicated using the Tabas accelerogram scaled to a 1 in 2,500 year event. There was a negative (i.e. upslope) residual movement indicated when modelling a 0.4 g event with the Tabas accelerogram as this accelerogram is asymmetric with larger pulses in one direction than the other.

6 Summary and Conclusions

Assessing likely lateral movements of the Pole 3 building during a very large earthquake was a challenging task due to factors such as:

- Variable fill strengths due to varying sources and the unknown method of infilling the gully in the past.
- The difficulty of accurately sampling and testing a fill containing a range of soil types and the uncertain changes in soil strength with depth.
- The strong 3D effects with one end of the building founded on rock and the other on 25 m of unengineered fill.
- High design PGAs which could be varying along the length of the building due to the varying depth to rock.
- The need to limit damage even at a 1 in 2,500 year earthquake event.

A range of analytical methods were used to give a range of potential lateral movements as shown in Table 2 below. Each method gave different displacements but all methods generally indicated that a deep failure surface close to the rock fill interface is the critical case. It was considered that the three dimensional analyses were the most appropriate method to calculate yield accelerations and the method of Bray and Travasarou (2007) the most appropriate method to calculate displacements. Deep seated movements were assessed as unlikely to exceed 30 mm to 350 mm under the 2,500 year earthquake event which is well within the capacity of the base isolators which can withstand up to 600 mm of lateral movement.

There are a number of factors that could affect this estimate but which could not be readily assessed. The presence of stronger partially saturated soils could reduce displacements along deep seated surfaces but multi-directional shaking and vertical accelerations could increase displacements. The gully fill is in the order of 150 m long and 35 m deep and it is likely that there will be out of phase inertial forces within the soil mass. Hence movements calculated assuming a uniform distribution of earthquake loading across the entire sliding mass is likely to be an upper bound value. Overall, the 30 mm to 350 mm estimated lateral displacement is considered to be a reasonable prediction of future behaviour.

Acknowledgments. I would like to acknowledge Transpower and Siemens who gave permission for this paper to be published. Thanks also go to Ian Brown for his help in 3d modelling, Dominic Mahoney who carried out the finite element modelling and Robert Kamuhangire for his editing prowess.

References

Ambraseys, N.N., Menu, J.M.: Earthquake induced ground displacements. Earthq. Eng. Struct. Dyn. **16**, 985–1006 (1988)

Bray, J.D., Travasarou, T.: Simplified procedure for estimating earthquake induced deviatoric slope displacements. J. Geotech. Geoenviron. Eng. **133**(4), 381–392 (2007). ASCE

GNS, McVerry, G.H.: Recommended accelerograms and vertical spectra for Haywards, GNS report 2007/362 (2007)

Kramer, S.L.: Geotechnical Earthquake Engineering. Prentice Hall, Upper Saddle River (1996)

NZGS/DSIR Beetham, R.D., Fellows, D.L.: Haywards site inspections – a seismic hazard and full stability evaluation, DSIR report EGI 87104 (1987)

Field Study on Response of Laterally Loaded Pile in Clayey Soil

S. Sivaraman(✉) and K. Muthukkumaran

NIT, Trichy, Tamil Nadu, India
sivaraman.geotech@gmail.com, kmk@nitt.edu

Abstract. Behavior of pile under lateral loading is a composite problem for soil-structure interaction. Field study data on such cases are not available in a large number, inspite of its essentiality due to site specific behavior. This paper deals with the field experimental study on behavior of laterally loaded pile in clayey soil. The site selected for the study is located near the Nagapatinam coastal line of Tamilnadu state of India. A series of laboratory tests and Standard Penetration Tests were conducted to characterize the soil. The lateral load tests were conducted on a individual bored pile which passes through soft to hard consistent clay upto a depth of 16 meters from the ground level. The pile was casted with the inclinometer casing to obtain the lateral displacement profile of pile. The displacement of pile was measured at the initial and final stage of lateral loading. Numerical studies were performed using the measured soil parameters. The static load-diaplacement response arrived as a result of field test were compared with the results of the numerical study.

1 Introduction

Deep foundation in the form of piles are generally adopted when the soil at the shallow depth has low bearing capacity or to transfer heavy loads to the deeper strata with high bearing capacity. The pile foundations are not only designed for the axial loads but also for the transverse loads in case of offshore marine structures, tall transmission towers, bridge foundations, etc. Many researchers have conducted a extensive studies on behavior of axially loaded piles tested both in lab and field. It is found that only a limited number of studies have been made to depict the lateral load behavior of pile in field. The design curves for predicting the deflection profile of pile were initailly developed by Matlock and Reese (1960). Field lateral load test on piles at various loactions were carried out by Alizadeh and Davisson (1970), Meyerhof et al. (1988) and Wang et al. (2002) and Brown et al. (1988). Based on the results of lateral load tests on steel pipe pile Zhou (1985) proposed D method. The present study was made to analyze the behavior of lateral loaded pile. The lateral pile load test for the present study was conducted at the large size grain storage godown construction site near Nagapattinam coastal region of tamil Nadu state. Since the site is located near the coastal region, the foundation for such structure needs to be designed for lateral load also. This paper discusses the response of the instrumented lateral pile load test and compares it with the results of numerical studies. Inclinometers are used to assess the

T. Abdoun and S. Elfass (eds.), *Soil Dynamics and Soil-Structure Interaction for Resilient Infrastructure*, Sustainable Civil Infrastructures, DOI 10.1007/978-3-319-63543-9_17

lateral deflection behavior of the piles subjected to lateral loads. In the present study, the pile was casted with inclinometer casing pipe tied along the reinforcement cage.

2 Site Geological Condition

The test site is located very close to the coastal line ie., site is about 2.5 km from the Bay of Bengal. The ground surface is found to be level and becomes water logged area during rainy seasons. The top soil of the site area is clayey loamy soil with very soft consistency. Several Standard Penetration Tests (SPT) were conducted at the site and the soil profile below the ground level is obtained as follows. The soil layer from the ground level upto a 5 m depth was found to be very soft low compressible clay with silts and sand. A very soft high compressible clay with silt and sand was existing from 5 m till 8 m. Below that a stiff to hard low compressible clayey soil with silt and sand was found upto 15 m depth. This layer was followed by a very dense clayey sandy layer even upto 21 m depth. Since the soil strata upto 15 depth was clayey soil with compressible nature, it is decided to rest the pile at a depth of 16 m in a very dense sandy layer. The soil strata around the pile is shown in Fig. 1. The variation of SPT ‘N’ values with respect to depth is represented in Fig. 2.

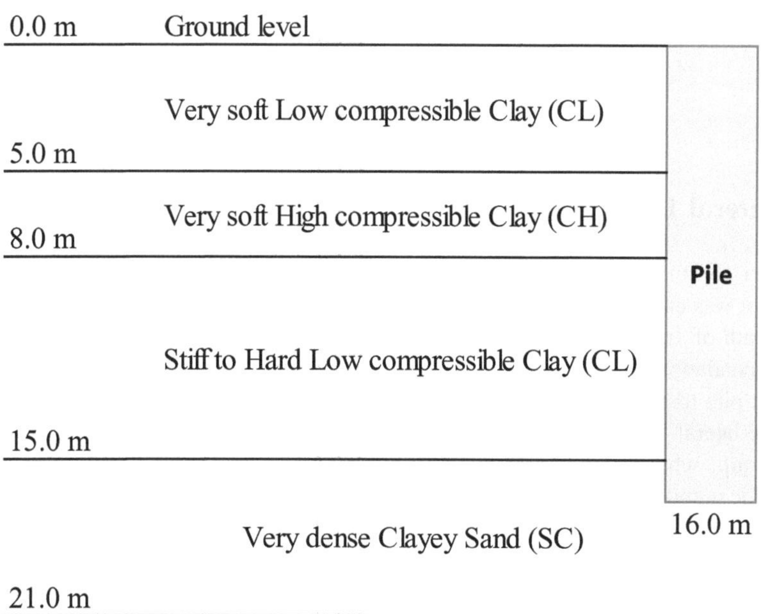

Fig. 1. Soil profile at the site of test pile

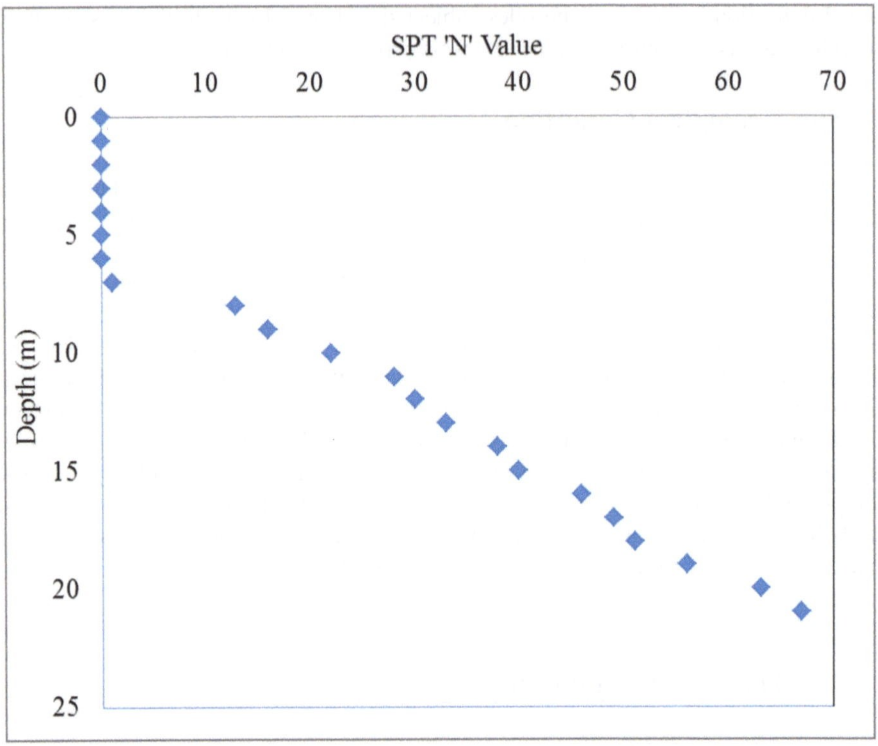

Fig. 2. Variation of SPT 'N' value with depth

3 Lateral Load Test

In order to study the lateral load behavior pile, a bored cast in situ pile of 0.6 m diameter was casted with inclinometer pipe incorporated in it. The pile was constructed to a depth of 16.0 m from the ground level. The pile was casted and tested as per the recommendations of IS 2911 Part-4. A linear three pile group was casted adjacent to the test pile to provide a reaction during lateral pile load test.

The lateral load to the pile was applied by a hydraulic jack acting on the reaction pile group, which was placed at the pile cut-off level. The jack is connected to a hydraulic pump. The load applied is calculated using the readings given by the pressure gauge. A dial gauge was installed diametrically opposite to the jack and deflection of the foundation is measured on its cut-off level. The load was applied in stages in accordance with IS 2911 Part-4 Sect. 7.1. The inclinometer readings were also recorded before and after the lateral load application to the pile.

4 Numerical Analysis

A numerical analysis was carried out using the fiinite element software PLAXIS 2D to compare its results with that of experiment. The general Mohr-Coulomb Model and Pile element were used to model the soil and pile for the analysis. The present problem is analysed as axis symmetric and to prevent boundary effect, the soil is modelled with an extend of 8 times the diameter of the pile. The 15-node fine mesh triangular elements were selected as they are effective choice for complicated problems. The sub soil properties used in the numerical analysis were assessed from the SPT 'N' values and the laboratory tests on disturbed and undisturbed soil samples. The soil parameters used in the numerical analysis are presented in Table 1. For the clayey soil, a fractional value of angle of internal friction is considered for better numerical analysis. The numerical model stimulated in the PLAXIS 2D software is shown in Fig. 3.

Table 1. Subsoil properties used for numerical analysis

Soil layer	Unit weight, γ (kN/m^3)	Young's Modulus E (kN/m^2)	Poissons ratio, ν	Cohesion C_u (kN/m^2)	Angle of internal friction, ϕ
CL	14	15000	0.35	20	0.001
CH	16	21000	0.35	70	0.001
CL	16	30000	0.33	105	0.001
SC	18	36000	0.3	10	32

5 Results and Discussions

The displacement profile at pile cutoff level for the apllied load values are graphically shown in Fig. 4. The values obtained from the inclinometer and the numerical analysis are found to be comparable. The load-displacement curve varies linearly upto maximum load and displacement increases rapidly after it.

It is also observed that the maximum lateral deflection at the maximum lateral load of 130 kN is found to be 3.19 mm.

6 Conclusion

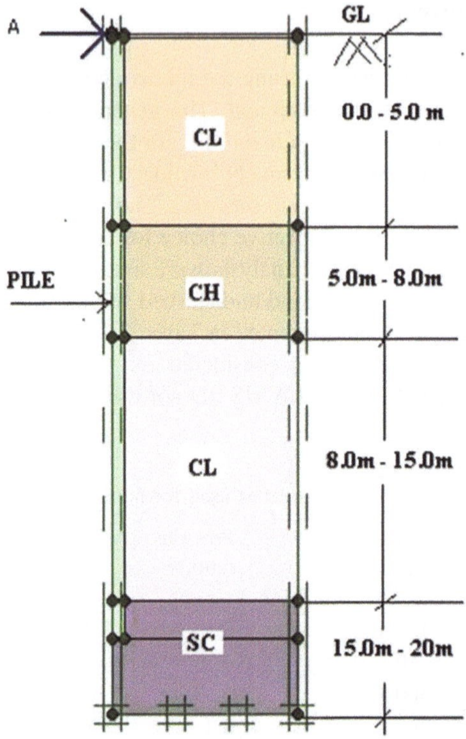

Fig. 3. Numerical soil model with applied load

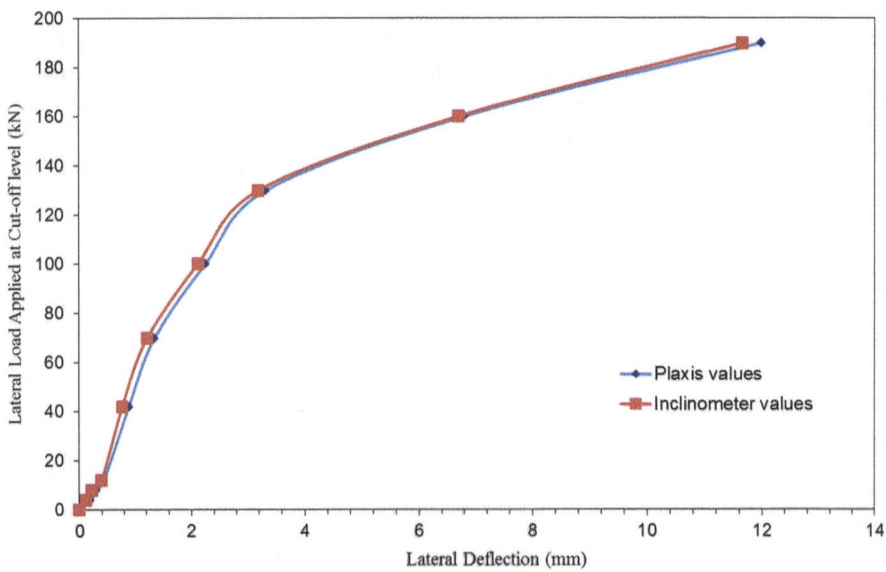

Fig. 4. Lateral Load vs displacement at cutoff level

The prediction of lateral load capacity of the pile plays a vital role in the design of foundation for tall buildings. In the present paper, the lateral load behavior of pile is assessed by conducting a lateral pile load test at a particular site. The present study is made as a part of pile design process and it is found to be a significant contribution. Further the numerical study using the subsoil parameters measured from lab and field tests were carried out. The results of such numerical study proves to be comparable with that of experimental results.

References

Alizadeh, M., Davisson, M.T.: Lateral load tests on piles-arkansas river project. J. Soil Mech. Found. Div. **96**(5), 1583–1604 (1970)

Brown, D.A., Morrison, C., Reese, L.C.: Lateral load behavior of pile group in sand. J. Geotech. Eng. **114**(11), 1261–1276 (1988)

IS 2911 Part IV: Code of Practice for the Design and Construction of Pile Foundations – Load Test on Piles, New Delhi

Matlock, H., Reese, L.C.: Generalized solutions for laterally loaded piles. J. Soil Mech. Found. Div. ASCE **86**(5), 63–91 (1960)

Meyerhof, G.G., Sastry, V., Yalcin, A.S.: Lateral resistance and deflection of flexible piles. Can. Geotech. J. **25**(3), 511–522 (1988)

Mei, W., Zhigang, L., Jiangxiang, L., et al.: Nonlinear analysis of m method for single pile under lateral loading. Rock and soil mechanics (china) **23**(1), 23–26 (2002). P-y

Ming, Z., Huichu, W., Ziai, L.: The analytical method for the experimental data of the laterally loaded piles in multiple layers of soil. Chinese J. Geotech. Eng. **7**(3), 25–40 (1985)

Effect of Backfilling Material Under Structures on Ground Motion Characteristics Due to Earthquake

Ahmed T.M. Farid(✉)

Geo-Institute of the HBRC, Dokki, Giza, Egypt
atfarid2013@gmail.com

Abstract. Due to limited areas and excessive cost of land for projects, backfilling process is become necessary. Also, backfilling will be done to overcome the un-leveling depths or raising levels of site construction, especially near the sea region. Therefore, backfilling soil materials used under the foundation of structures should be investigated regarding its effect on ground motion characteristics, especially at regions subjected to earthquakes. In this research, three thicknesses of sandy fill material of 20 m, 40 m and 60 m thicknesses are used to predict the modified ground motion characteristics effect at the foundation level. Comparison between the effect of using different thicknesses of sandy fill material above the natural soil strata on the recorded earthquake is studied, i.e., peak ground acceleration, time history, and spectra acceleration values. Shake computer program was used to perform this study. Strong earthquake records, with PGA (Peak Ground Acceleration) of (0.35 g) were used in the analysis. It was found that, higher fill material thickness has a significant effect on eliminating the earthquake ground motion properties at surface layer of fill material, near foundation level. It is recommended to consider the fill material thickness in the design of foundations subjected to seismic motions and more future studies should be analyzed for different fill and natural soil deposits.

1 Introduction

In many construction sites, leveling is necessary for establishing under different types of constructions. Most of leveling works are needed while there is soft soils or excessive cost of lands in a special region. Leveling is performed using backfilling material of different thicknesses and types. In region subjected to seismic motions from earthquakes, the using of fill material could effect on the seismic motions behavior. Fill material could change the spectral acceleration propagate from earthquake through soil layers till reach the foundation level. This spectral acceleration can be changed from site to site depending on the fill material thickness or property above the natural soil and bed rock formations. The present paper is attempts to study the effect of using backfill material of different thicknesses on the characteristics of the earthquake motions at foundation level. Three thicknesses of sandy fill material are used in our study having different thicknesses of 20 m, 40 m, and 60 m, separately. Fill material underlying by the natural clayey deposit above the rock formation. The sum of both of the fill material and natural soil deposit of about 300 m depth above bedrock formation.

T. Abdoun and S. Elfass (eds.), *Soil Dynamics and Soil-Structure Interaction for Resilient Infrastructure*, Sustainable Civil Infrastructures, DOI 10.1007/978-3-319-63543-9_18

Evaluation of site spectra and amplification influence for the different fill material thickness compared to the base rock stratum is studied. The effect of fill material thickness on the characteristics of a higher selected earthquake motion was carried out assuming a horizontally stratified soil deposits and vertically propagating seismic waves. The soil properties are varying with depths but remain constant in horizontal direction. Generation of acceleration time histories, frequency content, Spectral accelerations, and amplification factors related to bed rock are performed for each fill material thickness above bedrock, using SHAKE91 program. The control motions have been specified at the bedrock as datum for the analysis. The zero period ground acceleration for these earthquake is assumed as 0.10 g during this study.

The recording data at surface fill material normally reflects the seismic characteristics due to the fill material thickness of layers at the site. Therefore to obtain the response of any soil deposit, the following items should be studied: determination the fill material and natural soil properties and topography, determination of the underlying rock configuration, dynamic module and damping ratio properties of each of the fill material, natural soil and rock formations, the characteristics of earthquake motions.

2 Soil and Earthquake Records Used in the Study

2.1 Soil Surface Topography

To conduct our study, three thicknesses of fill material are used in the analysis. Three thicknesses of the compacted fill material are 20 m, 40 m, and 60 m separately, were investigated in the present research. The backfill material consists of sandy deposit compacted with different ratios from bottom to top layers (near foundation level). Compaction was higher and reached its maximum percentage value near the foundation level and was a fewer less with reaching the natural soil formation below. This was done to represent the actual case of execution any higher thickness of fill material at such situation. According to that different shear wave velocities were used to represent the actual compacted behavior of the sandy formation above the natural soil deposit. The shear wave velocities were ranged between 140 and 430 m/s for the compacted backfill material, where the higher value was near the foundation level and the lower one just when fill material reach the natural soil deposit. The depth of soil formation above bed rock was chosen to be about 300 m (includes both the backfill material and the natural soil deposit). Geotechnical data regards to the dynamic module and damping ratio properties of the soil formations are chosen from previous researches. Table 1 shows the dynamic module and damping ratio for fill material while, Table 2 shows the corresponding data for the natural clayey soil deposit above bed rock formation.

2.2 Earthquake Ground Motion Records

A strong earthquake ground motion of a peak acceleration of 0.35 g is used in the present study. The earthquake records are normalized to 0.1 g for comparison between

Table 1. Dynamic module and damping ratio of fill material (Sandy soil)

Modulus of sand		Damping of sand	
Strain	G/G max	Strain	Damping
0.0001	1.000	0.0001	0.240
0.0003	1.000	0.0003	0.420
0.0010	0.990	0.0010	0.800
0.0030	0.960	0.0030	1.400
0.0100	0.850	0.0100	2.800
0.0300	0.640	0.0300	5.100
0.1000	0.370	0.1000	9.800
0.3000	0.180	0.3000	15.50
1.0000	0.080	1.0000	21.00
3.0000	0.050	3.0000	25.10
10.0000	0.035	10.000	28.00

Table 2. Dynamic module and damping ratio of soil (Clayey deposit)

Modulus of sand		Damping of sand	
Strain	G/Gmax	Strain	Damping
0.003	1.000	0.003	1.600
0.004	0.990	0.004	1.800
0.005	0.985	0.005	2.100
0.006	0.980	0.006	2.300
0.007	0.970	0.007	2.400
0.008	0.965	0.008	2.700
0.009	0.960	0.009	3.000
0.010	0.955	0.010	3.700
0.020	0.905	0.020	4.200
0.030	0.850	0.030	4.600
0.040	0.815	0.040	5.000
0.060	0.750	0.060	5.200
0.080	0.710	0.080	5.700
0.100	0.670	0.100	6.100
0.200	0.565	0.200	8.000
0.300	0.480	0.300	9.200
0.500	0.385	0.400	10.100
0.600	0.350	0.500	10.900
0.800	0.300	0.700	12.200
1.000	0.250	1.000	13.500

the surface layer below the foundation and at the outcrop (bedrock layer). Then the spectral displacement curves are plotted for 5% damping. Table 3 shows the El-Centro (ELC-S60E) earthquake records used in the analysis.

Table 3. Characteristics of EL-centro earthquake ground motion

Date of earthquake	Peak acceleration (% of G)	Ground velocity (mm/sec)	Ground displacement (mm)	Duration time (sec)
05/08/40	0.348	334	124	55.76

3 Study and Analysis

Our study was performed to get the effect of site conditions of using different thicknesses of backfill material on seismic ground motions. This to predict how the waves from the underlying rock is changed due to the geometrical and geological structures of fill material characteristic and what will be the waves transmitted to the surface layers. The effect of backfill material thickness on the earthquake time history acceleration, frequency content, spectrum acceleration curves, amplification factors, and Fourier amplitude between bed rock and surface fill layer will be discussed in the following:

3.1 Earthquake Time History

The selected earthquake motion is used as input motion at the bed-rock formation. Then, the acceleration time histories are then computed at surface layer of each thickness layer of the fill material using shake program. Earthquake motion at bed rock (outcrop) illustrated from Fig. 1, while, Fig. 2 shows the varying time history acceleration of earthquake wave propagation at the surface layers of fill material (ground level under footings) for the three different thicknesses 20 m, 40 m, and 60 m, separately. Comparing Figs. 1 and 2, it is obvious that, the acceleration time histories had been filtered during passing through the natural soil deposit and the fill material till reach surface layer, i.e., they have smaller zero peak ground acceleration (ZPGA) and less time periods at surface layer of fill material compared to bed rock layer.

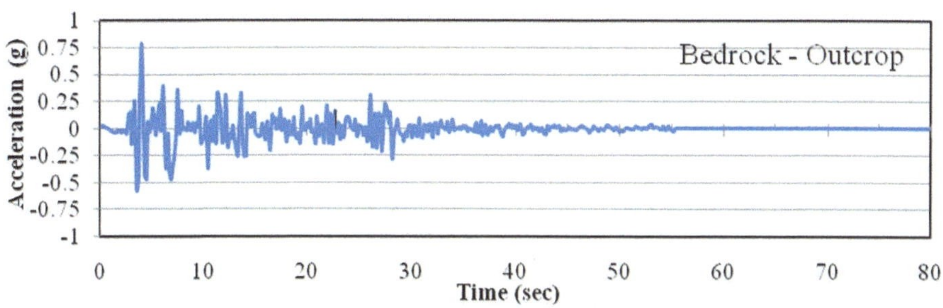

Fig. 1. Acceleration time history of earthquake at bedrock-outcrop

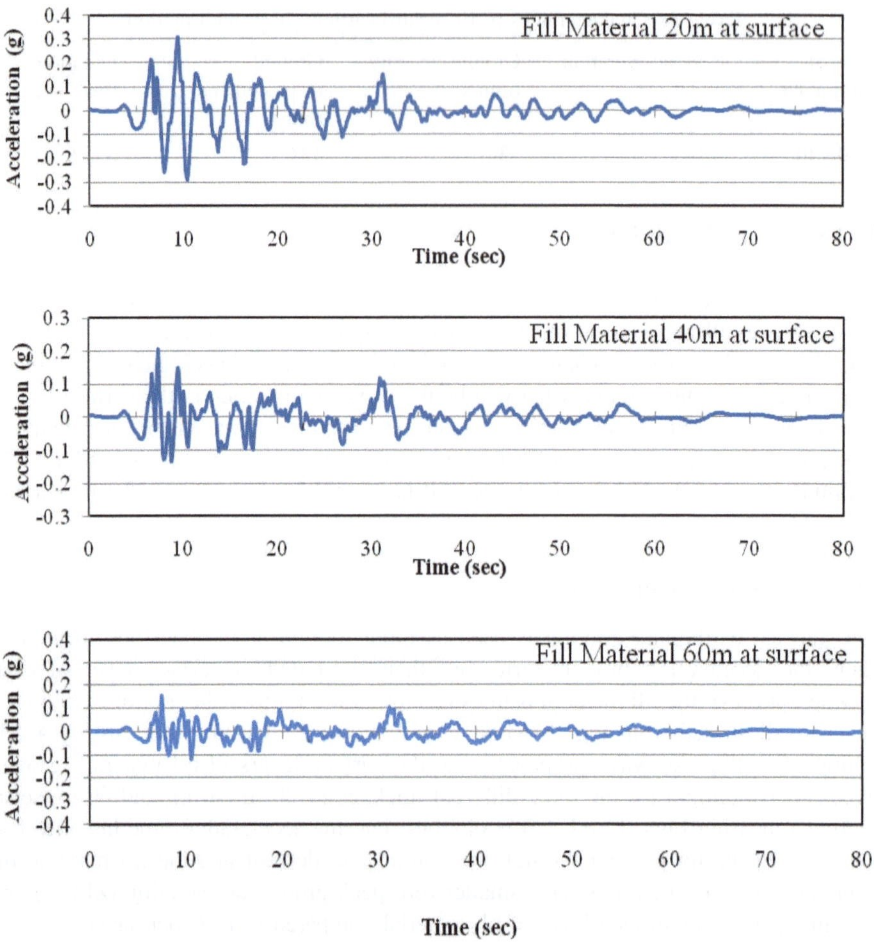

Fig. 2. Acceleration time histories for fill material of 20 m, 40 m and 60 m at Surface layer

3.2 Earthquake Response Spectrum

The response spectrums are plotted in Figs. 3-a, b and c for fill material of thicknesses 20 m, 40 m, and 60 m, respectively compared to bed rock. These figures give indication of the potential effects of the ground motions with the different thicknesses of the fill material. Comparing these three figures for the three fill thicknesses, it was illustrated that the acceleration at zero time periods was reduced with increasing the height of fill material with reducing the thickness of clayey soil deposit above rock formation. The reduction in the spectrum was ranged between 10%, 15%, and 25% for the three fill material thicknesses of 20 m, 40 m, and 60 m, respectively compared to the control

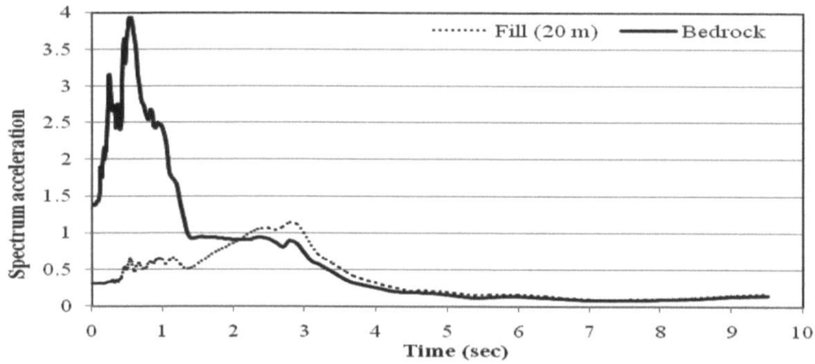

(a)

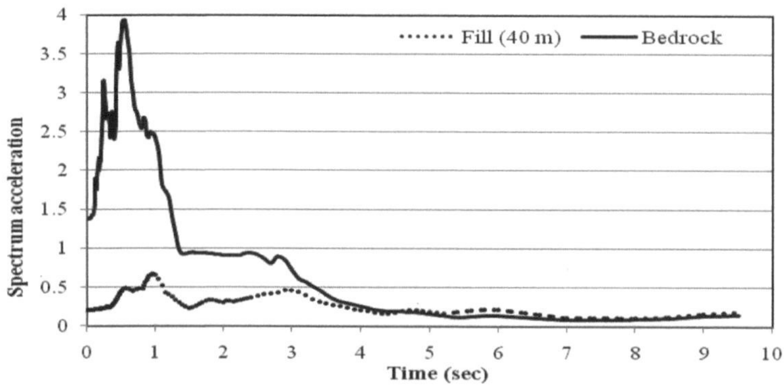

(b)

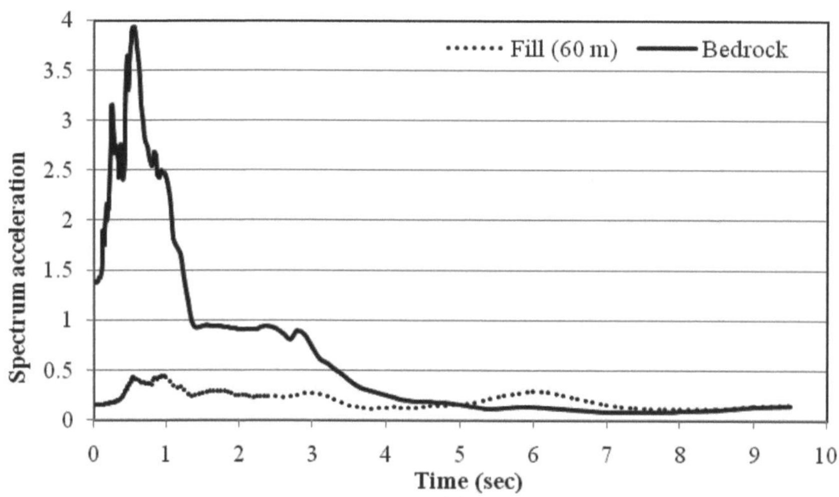

(c)

Fig. 3. (a) Comparison between acceleration response spectrums at surface of 20 m of fill material compared to bed rock formation. (b) Comparison between acceleration response spectrums at surface of 40 m fill material thickness compared to bed rock formation. (c) Comparison between acceleration response spectrums at surface of 60 m fill material thickness compared to bed rock formation

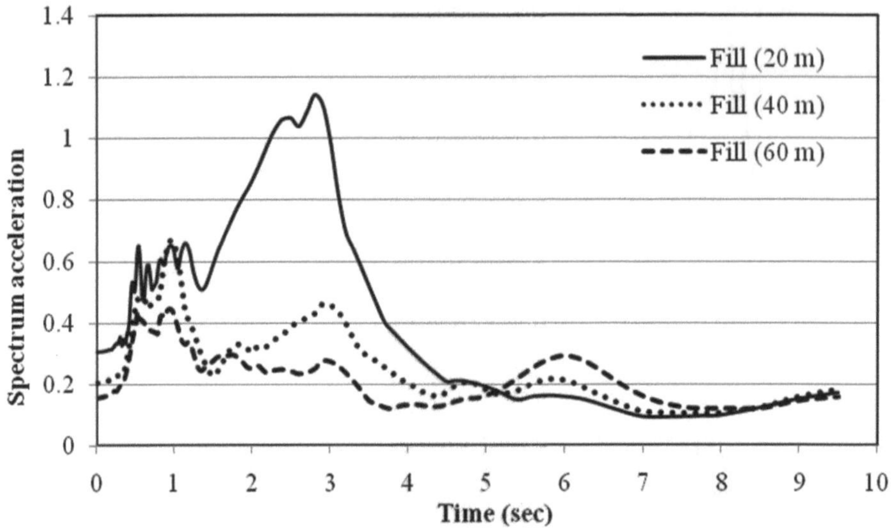

Fig. 4. Acceleration response spectrums of fill material with different layer thickness at surface

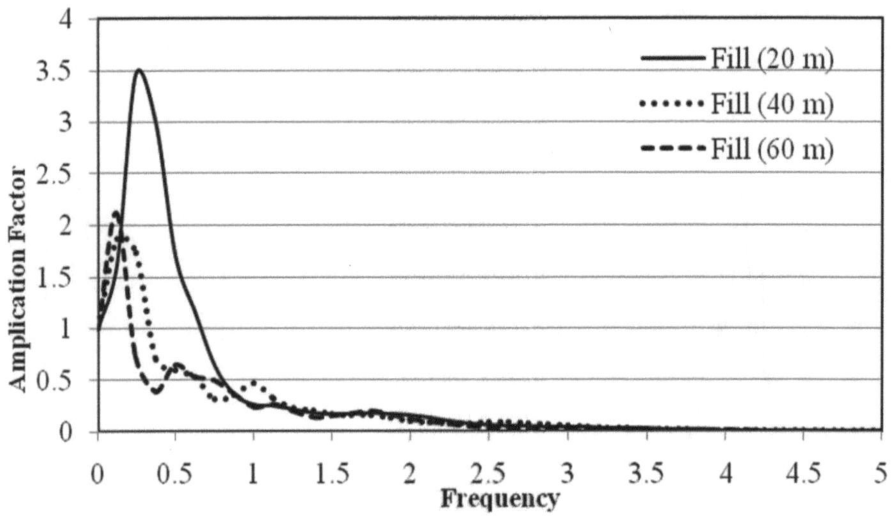

Fig. 5. Amplification spectrum between bed rock and fill material at surface for different Thickness

motion at the bedrock. Figure 4 illustrated the comparison between the reductions gained for the different fill material thicknesses. Amplification factors between surface layer and bed rock are founded in decreasing status with increasing the fill material thickness as shown in Fig. 5. This was due to increasing the clayey soil and increasing the sandy layer thickness as both thicknesses equal 300 m above rock formation.

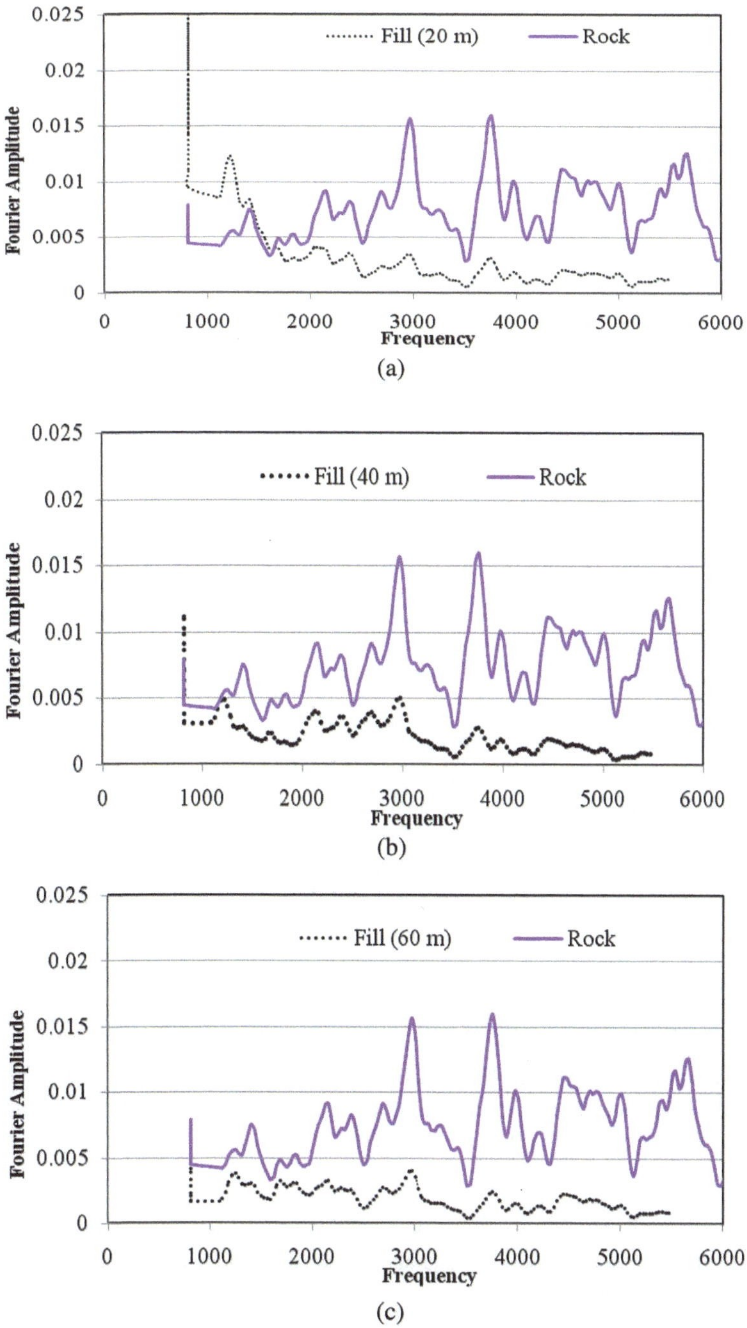

Fig. 6. (a) Fourier spectrum between bedrock and fill material of 20 m thickness. (b) Fourier spectrum between bed rock and fill material of 40 m thickness. (c) Fourier spectrum between bed rock and fill material of 60 m thickness.

3.3 Earthquake Fourier Spectrum

As shown in Figs. 6-a, b and c, the Fourier spectrum is a plot of amplitude versus frequency for fill material of 20 m, 40 m, and 60 m thicknesses, respectively with that for bed rock formation. The Fourier amplitude spectrum illustrates the frequency content of the motion. It is appeared that, there is a big difference in the amplitude by using fill material compared to bed rock data. Thus, using fill material will lower the earthquake Fourier spectrum amplitude and its frequency due to the far distance from rock seismic motion. It is obvious that, using of the fill material lower the frequency of seismic motion spectrum amplitude at surface layer.

4 Conclusions

The using of fill material for leveling above natural soil deposit underlain by bed rock formation for construction will effect on the seismic motion under buildings. The present study illustrated the following issues:

(1) The acceleration time histories was filtered during passing the natural soil deposit and then the fill material, i.e., they have a less time periods at the surface layer of the fill material with its thickness increase compared to bed rock layer time period.
(2) The reduction in the spectrum was ranged between 10%, 15%, and 25% for fill material thickness of 20 m, 40 m, and 60 m, respectively compared to the control motion at the bedrock.
(3) Amplification factors between surface layer of fill material and bed rock stratum decreased with increasing the fill material thickness due to the reduction of natural clayey deposit, as natural soil and fill material are of fixed thickness for all analysis equal to 300 m.
(4) Using the fill material will lower the earthquake Fourier spectrum amplitude and its frequency. According to that the fill material will lower the frequency of seismic motion spectrum amplitude at surface layer below foundation.

According to that, site characterization should be taken into consideration at seismic locations regions when design structures due earthquake effect and especially where fill material are used above natural soils. In these studies the behavior of seismic motions was affected by both the thickness of fill material and the natural soil formations above rock stratum. Recommendations for additional studies should be analyzed for different characteristics of fill material and natural soils thicknesses for further studies.

Acknowledgments. The author thanks Prof. Dr. Shamimul Rahman, North Carolina State University, USA for his collaboration and support for providing us the needed soft ware programs to perform the present study.

References

Charles, K., Pong, L.J.: Structural response amplification due to soft soil effect. Department of Civil Engineering, University of Toronto (2005)

Dobry, R., Vucetic, M.: Dynamic properties and seismic response of soft clay deposits. In: Proceeding, International Symposium on Geotechnical Engineering of Soft Soils, Mexico City, vol. 2, pp. 51–87 (2001)

Idriss, I.M., Sun, G.I.: SHAKE91: A Computer program for conducting equivalent linear seismic response analyses of horizontally layered soil deposit. University of California, Davis, California, pp. 13–18 (1992)

Macky, T.A.: Earthquake effect on response of soil deposit. Housing and Building Research Center, Egyptian Society for Soil Mechanics and Foundation Engineering, vol. 7 (1992)

Nawy, E.G.: SEISMOSIGNAL: A Computer program for earthquake characteristics. Version 3.1.0 (2005). Press Web Site at http://www.seismosoft.com/

ProShake User Manual: PROSHAKE: Ground response analysis program. Version 1.1. EduPro Civil Systems, Inc., Redmond, Washington (1996)

Sandford, T.C., et al.: Soil-structure interaction of buried structures. University of Maine, A2K04: Committee on Subsurface Soil Structure Interaction (2004)

Saxena, S.K., Avramidis, A.S., Reddy, K.R.: Dynamic moduli and damping ratios for cemented sands at low strains. Can. Geotech. J. **25**(2), 353–368 (1988)

Schnabel, P.B., et al.: SHAKE: A Computer Program for Earthquake Response Analysis of Horizontally Layered Sites. University of California, Berkeley (1972). EERC

School of Civil and Environmental Engineering: Ground motion amplification of soils in the Upper Mississippi embayment. Georgia Institute of Technology (2005). Press Web Site at http://www.ce.gatech.edu/

Seed, H.B., Idriss, I.M.: Soil moduli and damping factors for dynamic response analysis. University of California, Berkeley, California, EERC 70-10 (1970)

Soliman, K.Z., Farid, A.T.: Effect of clay deposit on ground motion characteristics and structure behavior. HBRC J. **2**, 96–105 (2006). Egypt

Sun, J.I., Golesorchi, R., Seed, B.H.: Dynamic moduli and damping ratios for cohesive soils. University of California Berkeley, EERC-88/15 (1988)

Zadoyan, P.: Site amplification effect at the Giumri industrial site. Nuclear and Radiation Safety Centre, civil Engineering, Ph.D. Dissertation (2000). Press Web Site at http://www.aaspe.am/

Guyed Monopile Foundation for off-Shore Wind Turbines

Reham M. Younis[1(✉)], Waleed E. El-Sekelly[2], and Ahmed E. El-Nimr[2]

[1] National Authority for Potable Water and Sewage, Giza, Egypt
rehamyounis3@gmail.com
[2] Mansoura University, Mansoura, Egypt
waleed.elsekelly@gmail.com, elnimrae@hotmail.com

Abstract. Over recent years, renewable energy has become an important choice for the Egyptian government, as Egypt has been suffering severe power shortages and rolling blackouts. Egypt put an energy strategy that aims at increasing the renewable energy share to 20 percent of Egypt's energy production by 2020. This target is expected to be met largely by scaling-up wind power projects. While most wind energy in Egypt is produced by on-shore wind farms, off-shore wind farms can be much more productive and efficient. Coastal zones in Egypt enjoy high wind energy potential. Designing foundations for off-shore wind turbines is challenging, as these structures are subjected to dynamic forces such as wind and waves. Monopiles are one of the most efficient foundation systems for offshore wind turbines, however they are mainly used in shallow water (less than 30 m deep). There is an urgent need to increase both the depth and capacity of monopiles supported wind turbines. Increasing the depth and diameter of monopiles can become uneconomic, thus a modified system needs to be adopted. This paper is the first step in studying the adoption of a system of ties (aka guys) to support the monopiles and reduce their free length above the sea level.

1 Introduction

Over recent years, renewable energy strategies resulted in great achievements in protecting the environment. Egypt tries to devote attention to the provision of energy through the adoption of renewable energy usage in order to assist in plugging its needs of electricity. This is due to the suitable climate and wide areas of land that are ready for wind projects at almost negligible coast. All over the world, offshore wind farms are expected to become an important source of energy in the near future to decrease dependence on fossil fuels. Europe has taken fast steps towards the development of offshore wind farms. It has begun rapid investment in the clean energy sector, with onshore and offshore wind power becoming larger part of that investment. As the innovation of wind turbines makes strides, the manufacturers have developed wind turbines with rotor diameters as large as 150 m and power ratings of over 7.5 MW to 10 MW. Recently, due to the expanding numbers of wind farms, it is planned to be positioned as far as 50 km from the shore and in water depths up to 50 m (Said et al. 2006; Abuzaid 2010; Malhotra 2011).

T. Abdoun and S. Elfass (eds.), *Soil Dynamics and Soil-Structure Interaction for Resilient Infrastructure*, Sustainable Civil Infrastructures, DOI 10.1007/978-3-319-63543-9_19

Loads applied to the wind turbines are generated from the combined effect of waves, rotating rotor blades and wind gust. To support these loads, special types of foundation need to be used. Guyed monopile is considered to be one of the most efficient and economical foundation alternatives. This paper is the first step for our research on guyed monopile. Guyed monopiles are used as a support structure for offshore wind turbine at deeper depths where using monopiles is economically inefficient.

Despite the potential benefits of using guys in reducing the cross sections and free length of the monopile shaft, little research about it can be found in the literature. This research aims at extensive analysis of guyed monopiles, using the software package (PLAXIS 3D). This paper is a literature review and the first step towards producing practical design charts for the analysis and design of guyed monopiles.

2 The Future of Offshore Wind Turbines in Egypt

Recently, the Egyptian Ministry of Electricity & Energy has embedded the renewable energy in its long term plan, as a result of the severe power shortages that Egypt faced over the past few years. The preference has been given to wind, solar and biomass energy sources. Wind energy utilization occupies the top of New and Renewable Sources of Energy (NRSE) preferences. Regions in Egypt enjoy high wind Energy potential, as Egypt has available wide and vacant land. Figure 1 shows data from the national wind resource estimation program based on measuring stations that confirmed the plentiful wind energy potential in the coastal zones (north coast, the Red sea coast of the Gulf of Suez, South Sinai and on the Nile banks in the eastern and western desert). The Egyptian government has recently established large scale Wind farms, such as Zafarana, where the wind speed can reach up to 8 to 9 m/s.

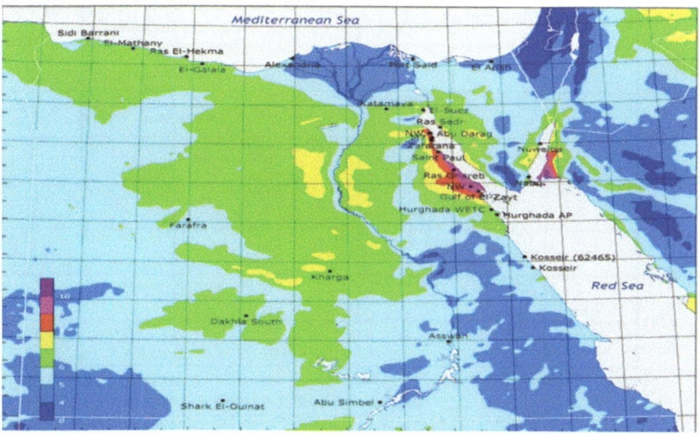

Fig. 1. Wind resource map of Egypt: mean wind speed at 50 m a.g.l, determined by mesoscale modeling (Said and Badger 2006).

While most wind energy in Egypt is provided by onshore wind farms, offshore wind farms can be far more productive and efficient. Figures 2 and 3 show places like (RasGhareb – Gulf of El Zayt) characterized by very high wind speed encouraging the construction of offshore wind turbines (Mostafa 2014; Hamouda 2011; Elsobki 2009; Said et al. 2006; Mortensen et al. 2006, 2003; Rizk 1987).

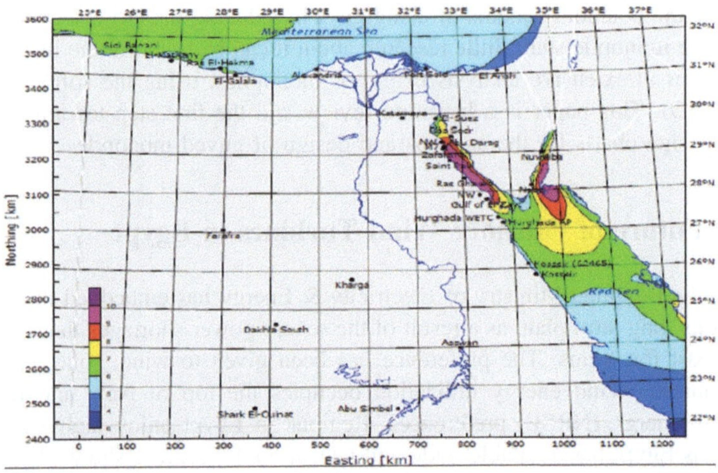

Fig. 2. Offshore wind resource map of Egypt: mean wind speed at 50 m a.g.l.determined by mesoscale modeling (Said and Badger 2006).

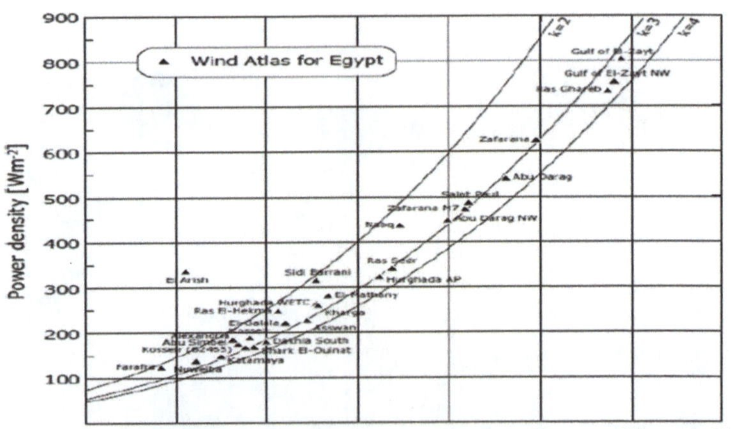

Fig. 3. Mean wind speeds and power densities at a height of 50 m over roughness class 1 (z0 = 0.03 m) for the 30 stations (Said and Badger 2006).

3 Benefits of Offshore Wind Turbine

Generally, the amount of power generated from wind turbines depends on wind speed, turbine size, and the arrangement of the turbines. Offshore wind speed is significantly higher than that of onshore, resulting in higher energy production per hour. Offshore wind turbines can harness the strong and steady ocean wind to produce clean electricity.

Various European and Asian countries have heavily utilized offshore wind farms in their energy production plan. Year after year, offshore wind energy utilization has been increasing, as shown in Fig. 4, suggesting the global awareness of the benefits and importance of offshore wind turbines, as an efficient renewable energy source.

(Cabral 2015; Maier 2015; Abuzaid 2010; Westgate and De Jong 2005)

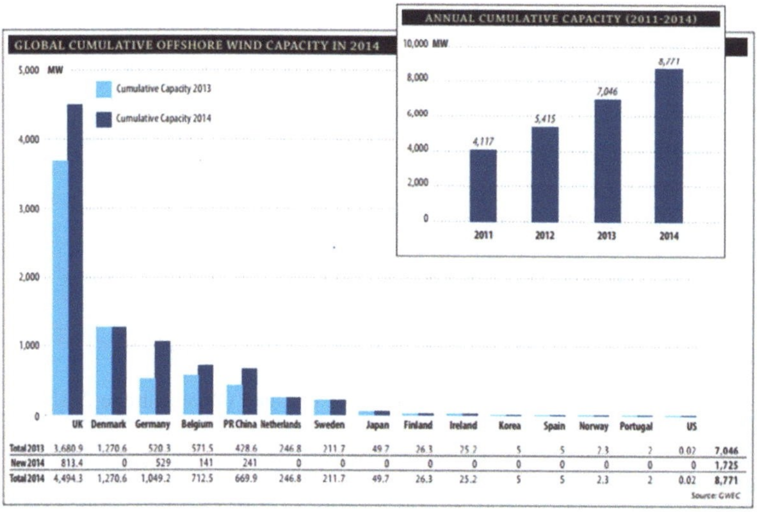

Fig. 4. Global offshore wind energy statistics. (Cabral 2015)

4 Planning the European Offshore Wind Industry 2016

The European offshore wind industry is working hard to expand this new clean technology in all European countries, as offshore wind energy is considered an ideal source of local, clean, and reliable energy.

Quoting from (wind Europe 2016) "114 wind turbines had an entirely grid connected, totaling 511 MW in 4 wind farms: Westermeerwind (NL), Gemini (NL), Gode Wind I (DE), Gode Wind II (DE). 182 turbines (44 units or 32% more than during the same period last year) had constructed in four wind farms:Westermeerwind (NL), Gemini (NL), Gode Wind I (DE), Gode Wind II (DE). 177 foundations (28% more than the same period last year) had established in nine wind farms: Nordsee One (DE),

Nobelwind (BE), Nordergründe (DE), Dudgeon East (UK), Sandbank (DE), Veja Mate (DE), Burbo Bank Extension (UK), Race Bank (UK), Rampion (UK). Piling activity at Wikinger (DE). The average size of wind turbines placed in the first half of 2016 is 4.8 MW, or 15% larger than over the same period last year. Seven projects, meriting €14bn, reached Final Investment Decision (FID) in the first half of 2016. This will finance 3.7 GW of new capacity, a doubling from the first half of 2015 (1.8 GW). Cumulatively, there are 3,344 offshore wind turbines with a combined capacity of 11,538 MW fully grid connected in European waters in 82 wind farms across 11 countries, including demonstration sites". Due to the wide spread of offshore wind turbines, it has occupied the top of efficient installation techniques, as shown in Fig. 5.

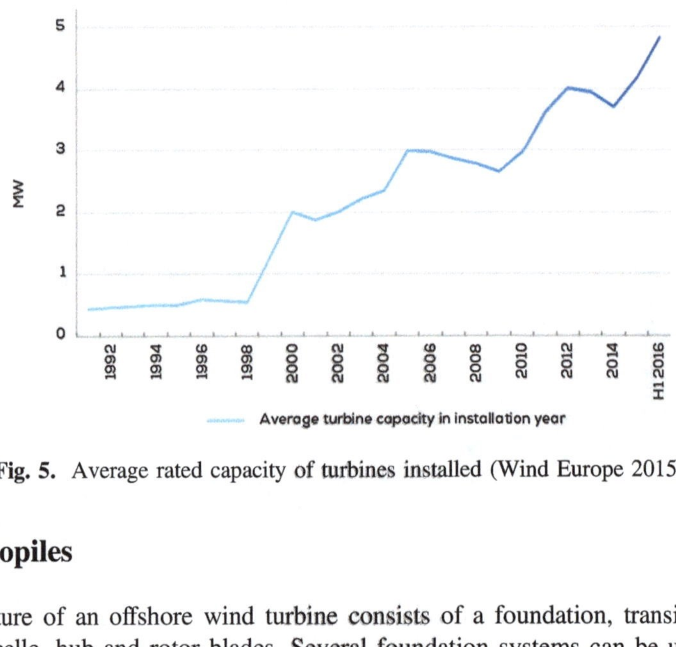

Fig. 5. Average rated capacity of turbines installed (Wind Europe 2015).

5 Monopiles

The structure of an offshore wind turbine consists of a foundation, transition piece, tower, nacelle, hub and rotor blades. Several foundation systems can be used in offshore wind turbines, such as tripods, jackets, gravity foundations, monopiles and floating foundation, as shown Fig. 6. Monopiles are the most popular foundation system used in the offshore wind energy sector. Monopiles are hollow steel piles with diameters ranging from 4 to 6 m and wall thickness from 40 to 80 mm typically used in shallow water, up to 30 m, and are driven or drilled into the soil.

By the end of 2015, approximately 80.1% of the foundations of offshore wind farms were monopiles, 9.1% gravity structures, 5.4% jackets and 3.6% were tripods as shown in Fig. 7.

Monopiles are easy to manufacture and transport, as compared to other types of foundation systems. It has been used for many years and considered to be the most cost effective and practical form of foundations, especially for shallow water depths. The advantages of utilizing monopiles include: minimal seabed preparation requirements,

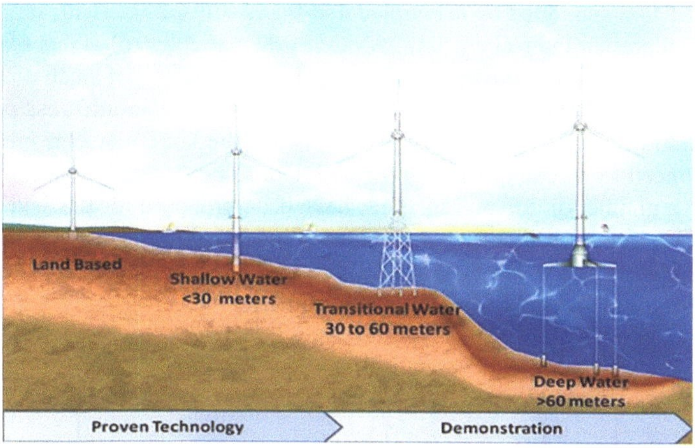

Fig. 6. Offshore wind turbine foundations. (Image courtesy of national renewable energy lab)

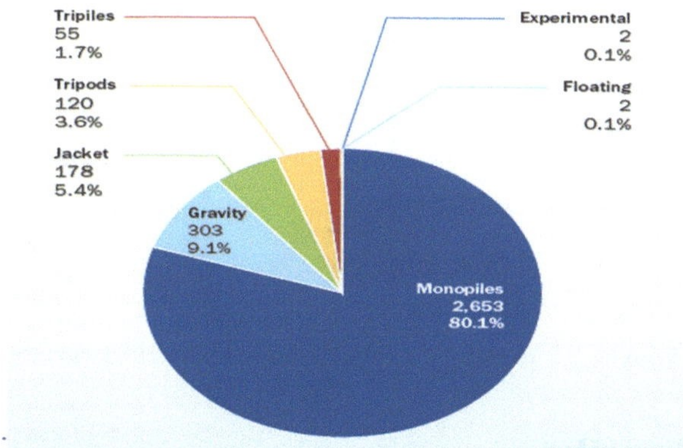

Fig. 7. Substructure types for online wind energy end of 2015 (EWEA 2015 Europe).

resistance to seabed movement, scour, ice flow damage and reasonable production costs. (Alejandra 2015; Kallehave et al. 2015; Maier 2015; Carswell 2012; Malhotra 2011; Saleem 2011; Bekken 2009; Zaaijer 2003)

6 Methods of Monopile Analysis

Monopiles are traditionally analyzed according to conventional methods and standards for laterally loaded piles, such as the American Petroleum Institute (API) and a number of versions of p-y curves. Quoting from Haiderali et al. (2013) "The p-y method is a subgrade reaction technique that describes the non-linear relationship between the soil resistance mobilized against the pile,(p), and the lateral deflection of the pile,(y) (Brødbaek et al. 2009). In this method, the pile is considered as a Winkler beam on the elastic foundation supported by uncoupled non-linear soil springs (with spring stiffness E_{PY}) which represent the soil reaction. A plot of the variables (p) and (y) at a discrete soil depth constitutes a p-y curve", as shown in Fig. 8. Analysis of small diameter piles can be safely performed by the p-y method that takes the different stress components indirectly into account. However the p-y curves for the large diameter piles (such as monopiles) need to be extrapolated and was found to be generally inaccurate. On the other hand, a number of different methods have been proposed for the development of the p-y curves for large diameter monopiles using 3D finite element codes, such as Plaxis 3D (Cabral 2015; Da Fonseca 2015; Leite 2015; Haiderali and Madabhushi 2012; Doherty and Gavin 2011; Martínez-Chaluisant 2011)

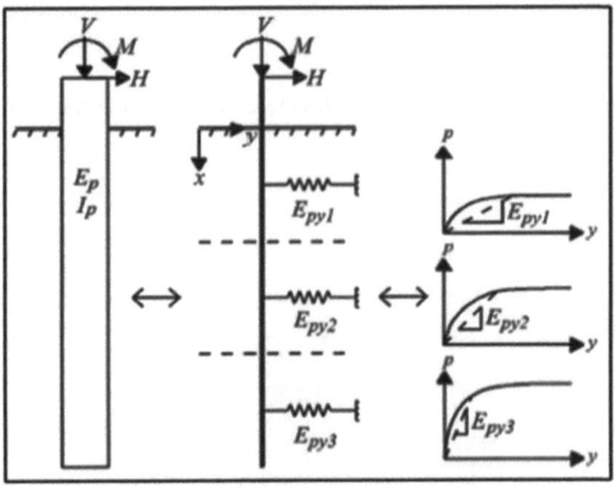

Fig. 8. Winkler p-y model for a laterally loaded pile (Haiderali et al. 2013)

7 Guyed Monopiles

While monopiles may be the ideal choice for shallow water depths, less than 30 m, it may be uneconomical for deeper water. At transitional water depth (30 to 50 m), the hydrodynamic loads and bending moments at the seabed increase, and the dynamic behavior of the structure changes. Finite elements analysis of monopiles, performed by several researchers showed large lateral deflection occurring as a result of large lateral loads, as shown in Figs. 9 and 10 (Bekken 2009). In order to overcome this problem at larger water depth, researchers and practitioners increased the monopile diameter and wall thickness, resulting in a huge bulk of steel which is heavy and expensive to install. Hence, monopiles were not found to be best solution economically at greater depths. Traditionally, the foundations used at these depths are gravity base structures, tripods and jacket towers as shown in Fig. 11. Unfortunately, these structures are heavy, and very expensive to fabricate and install. Using a system of ties to support monopiles (guyed monopiles) may be an efficient solution in reducing the cross section of monopiles, as it reduces the free length and thus the straining actions of the monopiles (Fig. 12).

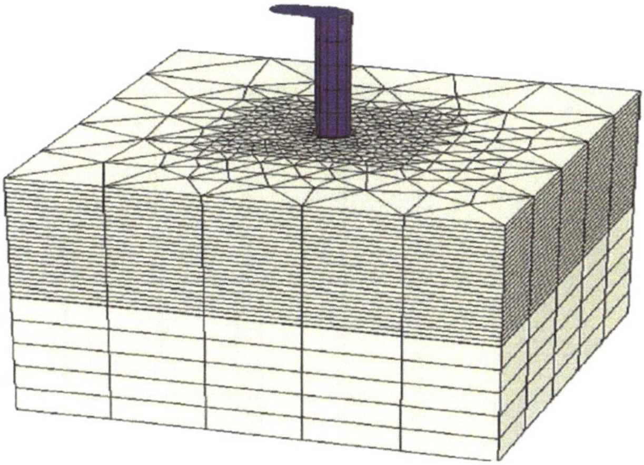

Fig. 9. Analysis of a monopile foundation for offshore wind turbines on PLAXIS 3D (Bekken 2009)

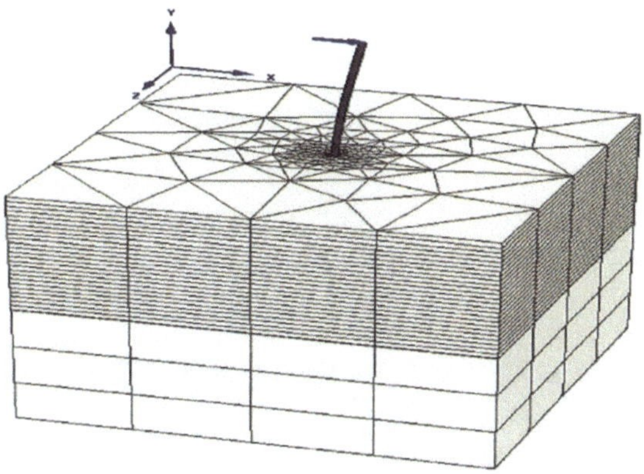

Fig. 10. Show deflection of monopile that occurs after run on PLAXIS 3D (Bekken 2009)

Despite the potential benefits of using guys in reducing the cross sections and free length of the monopile shaft, little research discussing this topic can be found in the literature. In our research, we aim at performing extensive finite elements analyses (using plaxis 3D package) on guyed monopiles in order to produce design charts that can be easily used by practitioners is the field. (Aspizua 2015; Kallehave et al. 2015; Maier 2015; Bhattacharya 2014; Fai and Khow 2008; Carey 2002)

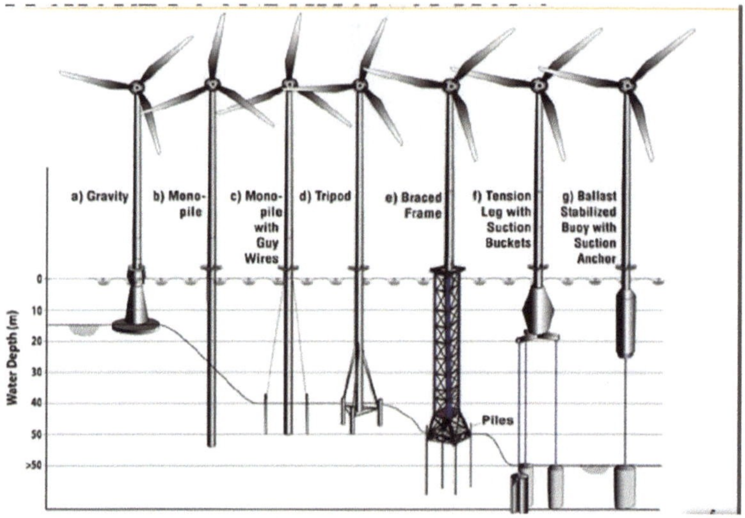

Fig. 11. Various types of support structures and their applicable water depth (Malhotra 2007).

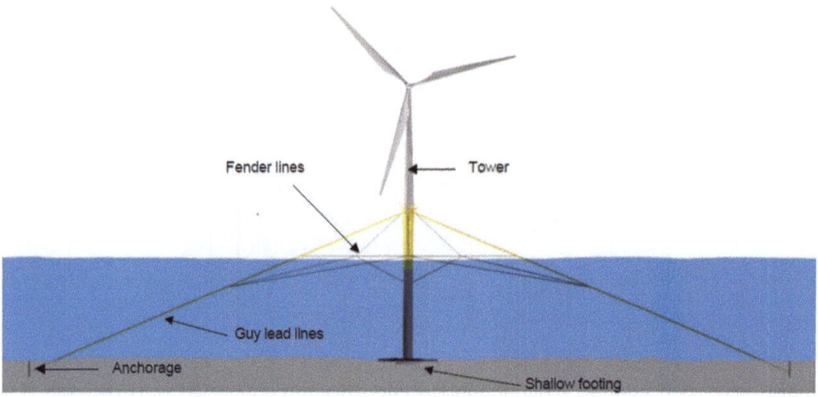

Fig. 12. Model of guyed monopole (Carey 2002)

8 Conclusions

Over the last few decades, offshore wind turbines are considered ideal sources to generate clean energy. Offshore wind-turbines can harness the strong and steady ocean wind which can help in generating clean electricity.

The structural foundation systems for offshore wind turbines are challenging, especially at water depths greater than 30 m. Monopile foundation is not considered the best solution at water depths beyond 30 m. Instead, Guyed monopile could be one of the most cost- effective foundation systems. Guyed monopiles structures may be considerably light and economical as compared to other foundation alternatives. Extensive research is currently under development by our team to study the behavior of monopiles and to generate design charts that can be easily used by practitioners is the field.

References

Abuzaid, F.: Offshore wind energy. PH240, Stanford University (2010)

Alejandra, M.: Centrifuge evaluation of offshore monopile systems. Rensselaer Polytechnic Institute, Master thesis, Troy, New York (2015)

Aspizua, L.: Offshore foundation-a challenge in the Baltic sea. Master thesis, Denmark (2015)

Bekken, L.: Lateral behavior of large diameter offshore monopile foundations for wind turbines. Doctoral dissertation, TU Delft, Delft University of Technology (2009)

Bhattacharya, S.: Challenges in design of foundations for offshore wind turbines. J. Eng. Technol. Ref. **1** (2014)

Brødbaek, K.T., Møller, M., Sørensen, S.P.H., Augustesen, A.H.: Review of p-y relationships in cohesionless soil, DCE Technical report No. 57, Department of Civil Engineering, Aalborg University, Aalborg, Denmark (2009)

Cabral, A.: Diameter effects of large scale monopiles. Master thesis in University of Portugal, Portugal (2015)

Carey, J.: Guyed support structures for offshore wind turbines. In: Proceedings of Gigawind Symposium, Germany (2002)

Carswell, W.: Probabilistic analysis of offshore wind turbine soil-structure interaction. Doctoral dissertation, University of Massachusetts Amherst (2012)

Da Fonseca, A.C.V.: Diameter effects of large scale monopiles-a theoretical and numerical investigation of the soil-pile interaction response. Master thesis (2015)

Doherty, P., Gavin, K.: Laterally loaded monopile design for offshore wind farms (2011)

Elsobki, M.S.: Wind Energy in Egypt. Industrial Modernization Program (2009). http://www.wto.org/english/tratop_e/envir_e/wksp_goods_sept09_e/elsobki_e.pdf

EWEA: The European offshore wind industry key trends and statistics 1st half 2016. European Wind Energy Association (EWEA), 1, 8 (2016)

EWEA: The European offshore wind industry key trends and statistics 1st half 2015. European Wind Energy Association (EWEA), 1, 8 (2015)

Fai, L.C., Khow, C.Y.: Offshore foundations for windfarms, Master thesis (2008)

Haiderali, A., Cilingir, U., Madabhushi, G.: Lateral and axial capacity of monopiles for offshore wind turbines. Indian Geotech. J. **43**(3), 181–194 (2013)

Haiderali, A., Madabhushi, G.: Three-dimensional finite element modelling of monopiles for offshore wind turbines. In: Proceedings of the World Congress on Advances in Civil, Environmental, and Materials Research, Seoul, pp. 3277–3295 (2012)

Hamouda, Y.: Wind Energy in Egypt. Working Paper Series. German University in Cairo Faculty of Management Technology Al Tagamoa Al Khames, 11835, New Cairo – Egypt (2011)

Kallehave, D., Byrne, B.W., Thilsted, C.L., Mikkelsen, K.K.: Optimization of monopiles for offshore wind turbines. Philos. Trans. R. Soc. A **373**(2035), 20140100 (2015)

Leite, O.B.: Review of design procedures for monopile offshore wind structures. Master thesis (2015)

Maier, S.: Feasiblity of offshore wind substructures in arctic environments (Doctoral dissertation, TU Delft) (2015)

Malhotra, S.: Selection, design and construction of offshore wind turbine foundations. INTECH Open Access Publisher (2011)

Malhotra, S.: Design and construction considerations for offshore wind turbine foundations. In: ASME 2007 26th International Conference on Offshore Mechanics and Arctic Engineering, pp. 635–647. American Society of Mechanical Engineers (2007)

Martínez-Chaluisant, V.: Static and dynamic response of monopiles for offshore wind turbines (Doctoral dissertation, University of Wisconsin-Madison) (2011)

Mortensen, N.G., Hansen, J.C., Badger, J., Jorgensen, B.H., Hasager, C.B., Paulsen, U.S., Moussa, A.S.: Wind atlas for Egypt: measurements, micro-and mesoscale modelling. In: 2006 European Wind Energy Conference and Exhibition (2006)

Mortensen, N.G., Said, U.S., Frank, H.P., Georgy, L., Hasager, C.B., Akmal, M., Salam, A.A.: Wind Atlas for the Gulf of Suez. Measurements and Modelling 1991–2001 (2003)

Mostafa, M.: Challenges to Energy Transition in Egypt: A Study of Wind and Solar Sectors (Doctoral dissertation, University of Potsdam, Germany) (2014)

Rizk, M.: Wind characteristics and the available wind energy in Egypt. Sol. Wind Technol. **4**(4), 491–499 (1987)

Said, U.S., Badger, J.: Wind atlas of Egypt. In: Mortensen, N.G. (ed.). RISO National Laboratory, New and Renewable Energy Authority, Egyptian Meteorological Authority, p. 12 (2006)

Saleem, Z.: Alternatives and modifications of Monopile foundation or its installation technique for noise mitigation. Report by Delft University of Technology for Stichting De Noordzee (the North Sea Foundation) (2011)

Westgate, Z.J., De Jong, J.T.: Geotechnical considerations for offshore wind turbines, Master thesis. Amherst, MA: University of Massachusetts and the Massachusetts Technology Collaborative (2005)

Zaaijer, M.B.: Comparison of monopile, tripod, suction bucket and gravity base design for a 6 MW turbine. OWEMES, ENEA, Naples (2003)

Dynamic Behavior of Marine Clay Soil Improved by Vertical Sand Drain

Manivannan Vinothkumar(✉) and K. Muthukkumaran

Department of Civil Engineering, NIT, Trichy 620015, India
mvkumar.c@gmail.com

Abstract. Numerical, analytical and experimental studies have shown that sand drain accelerates the rate of consolidation soft clay soil by providing the drainage path, reduction in settlement, increasing bearing capacity of the soil, this paper studies influence of cyclic behavior of sand drain imposed clay soil in laboratory cyclic Tri-axial apparatus, An experimental study on the consolidate drained dynamic behavior of saturated clays in cyclic triaxial tests with a variable confining pressure is presented. Tests were conducted on artificially prepared samples of marine clay (collected from Nagapattinam), using a dynamic triaxial apparatus where the deviatoric stress and confining pressure can be varied simultaneously. cyclic loading is applied with frequency of 0.25 Hz, drained cyclic loading the developed pore pressure was allowed to dissipate, virgin clay soil alone and improved with sand drained specimens having a maximum size of 100 mm in diameter and 200 mm in height was used for the study, Test results obtained from this study show that the drained response of saturated clays is strongly influenced by the application of sand drain, in terms of pore water pressure, development speed of cyclic strain and magnitude of cyclic strength, drainage property is compared with without sand drain specimens. Studies include effect of dissipation of excess pore water pressure and cyclic strain ratio.

1 Introduction

Soils are naturally accessible material and their properties are extremely passionate about the environmental condition in which they're deposit/formed.

The submarine condition extremely influences the design of any structure because it bears the load transfer from the structure. The present soil at given site might not be suitable to support the structures (Example: buildings, bridges, dams) due to low bearing capacity and Settlement problem. However the Urbanization and insufficiency of land lead to the compulsion to use the problematic surface area, like Soft clay soil, thence Ground improvement techniques are participating within the primary role in strengthening that specific weak soil deposit. Technique of ground improvement can have an intensive impact on the project span, fund provision, yet as performance of the super structure. Construction in coastal regions with deep deposit of soft sentimental clay is problematic as a results of excessive settlements and expansive nature of clay soil. Among varied technologies enforced by engineers. Vertical sand drain (VsD) is in every one of the leading adaptable ways that of improvement of sentimental soil, it

T. Abdoun and S. Elfass (eds.), *Soil Dynamics and Soil-Structure Interaction for Resilient Infrastructure*, Sustainable Civil Infrastructures, DOI 10.1007/978-3-319-63543-9_20

offers a permanent resolution in reducing post-construction settlement and for achieving faster rate of each radial and vertical drainage.

1.1 Proposed Project Using VsD

Ministry of Indian Railways, Southern division has scheduled to determine a new broad gauge Railway line between Nagapattinam and Thiruthuraipoondi, over an embankment that is found within the coastal region of the Bay of Bengal, These coastal regions are characterized by very weak and extremely compressible soft clay deposits, with low bearing capacity, low shear strength and extremely compressibility. Construction in such sub soil condition without soil improvement is usually imprecise as a result of the anticipated large and long-term settlement on the other hand, improvement soft ground is also found to be crucial when the sub soil low permeable soil, As the consolidation process occurs very slowly and takes many years to stabilize the settlement.

The site is found within the Nagapattinam District of Tamil Nadu, which lies on the costal line of Bay of Bengal, it is situated at 326 km south of the state capital, Chennai, and 145 km from Trichy a middle town. The Nagapattinam town lies between Northern Latitude 10.79° and 79.85° eastern longitude. The foundation soil in this region is found to be categorized by the existence of thick layers of soft marine clay deposits. Indian Southern Railways planned to determine a new broad gauge (BG) line between Nagapattinam and Thiruthuraipoondi.

Soil Profile

The ground water table lies at a depth of 3.0 m from the ground surface. The subsoil conditions ascertained from the borehole data are as follows:

I. Standard penetration test (SPT) observed 'N' Values varies from 5 to 13.
II. Undisturbed core samples (UDS) were test for its unconfined compression strength (UCS) and it was found to be 24–72 kN/m^2
III. Subsoil consists of very soft blackish marine clay having a natural water content (W) of 20–30% from ground level to 10 m depth.
IV. Atterberg's limits i.e., Liquid Limit (LL), Plastic Limit (PL), Shrinkage limit (SL) were lies between 50–60%, 20–25%, and 5–8% respectively throughout the depth.
V. Bulk unit density (γ_d) of the subsoil deposit were 15–17 kN/m^3.

Recommendations provided for the proposed project:

Based on the Barron's theoretical analysis and FEM study using Plaxis software, we concluded that 300 mm VsD with triangular spacing of 3.5 m center to center was adopted.

2 Experimental Programme

Soil samples from propose site were used to study the cyclic behavior with and without VsD, Single VsD therewith influence zone were scaled to 100 mm diameter and 200 mm diameter height sample, For sand drain around 100% replacement are used to for simulation.

Sample preparation the experiments were conducted on reconstituted samples prepared within the laboratory. The details of experiments conducted are elaborated below.

The experiments were carried out on normally consolidated samples, reconstituted from slurry by one-dimensional consolidation. This methodology of sample preparation closely resembles sedimentary deposits of clay within the field. Slurry was prepared with soil at a water content of 1.5 times the liquid limit. Shown in Fig. 1. The slurry was mixed thoroughly in a mortar mixer for enough time without clay lumps, Procter mould was used to prepare sample, and also the aspects of the mould were unclean with silicone grease to reduce the side friction. When placing the porous stone and filter paper separator at the bottom, the slurry was carefully poured in to the mould. Tamping was done to eliminate air voids while placing the slurry in the mould. A filter paper separator followed by a porous stone was placed above the slurry. When placing the loading cap above the porous stone, the sample was allowed to consolidate under two way drainage. Initially the sample was allowed to consolidate under the load of the loading cap for 24 h. Later it absolutely was transferred to a pneumatic loading frame and loads were applied in steps till a consolidation pressure of 100 kPa was reached. Minimum consolidation pressure of 100 kPa was found necessary so as to gain enough strength of the sample for handling. Sufficient time was allowed between each load in order that the clay didn't squeeze out at the time of loading. Under the ultimate stress level of 100 kPa, the sample was allowed for full consolidation. Once the consolidation was over, that was confirmed from the time-settlement curve, six thin walled sampling tubes of inner diameter of 100 mm was pushed into the mould. Samples was ejected out from the sampling tubes and trimmed off to a height of 200 mm to perform the tests.

3 Test Procedure

The following tests procedure were conducted on samples prepared from slurry consolidation technique to study the cyclic behaviour of consolidated clays and its comparisons. After the sample extract from the mould, sample is fixed in the cyclic triaxial base, saturation has been done by applying back pressure and confining pressure by gradually (difference in pressure was 0.5 kg/cm^2), pressures were applied (as per Skempton's parameters) until the ratio between the change in pore pressure to change in confining pressure reaches 95%, then the sample is allowed for consolidation process, sample without drains taken more time period then the sample with single VsD. Cyclic load is applied with frequency of 0.25 Hz., and the number of cyclic(N) was taken as 100.

Fig. 1. Typical setup for slurry consolidation

4 Cyclic-Strain Approach

The cyclic stress ratio, CSR, as proposed by Seed and Idriss (1971), this analysis was directed for the most part toward analysis of the loading conditions required to trigger liquefaction. This loading was delineated in terms of cyclic shear stresses, and liquefaction potential was evaluated on the idea of the amplitude and range of cycles of earthquake-induced shear stress. The overall approach has come back to be called the cyclic stress approach. Seed and Lee (1966) outlined initial liquefaction because the purpose at that the increase in pore pressure is equal to the initial effective confining pressure i.e., once (or once. as a result of most of the early laboratory testing investigations were based on cyclic triaxial tests on isotropically consolidated specimen

Table 1. Summary of dynamic behavior of clay sample with and without VsD

N	Max excessive pore water pressure		Cyclic stress ratio		N	Max Excessive pore water pressure		Cyclic stress ratio	
	Without VsD	With VsD	Without VsD	With VsD		Without VsD	With VsD	Without VsD	With VsD
1	0.18	0.121	−3.62	−2.534	21	0.18	0.125	−2.76	−1.932
2	0.24	0.165	−3.26	−2.282	22	0.18	0.127	−2.71	−1.897
3	0.21	0.145	−3.17	−2.219	23	0.18	0.128	−2.73	−1.911
4	0.18	0.125	−3.13	−2.191	24	0.18	0.125	−2.73	−1.911
5	0.2	0.144	−3.09	−2.163	25	0.17	0.125	−2.73	−1.911
6	0.2	0.145	−3.06	−2.142	26	0.18	0.125	−2.71	−1.897
7	0.19	0.135	−2.98	−2.086	27	0.18	0.111	−2.73	−1.911
8	0.19	0.136	−2.91	−2.037	28	0.17	0.12	−2.69	−1.883
9	0.18	0.124	−2.93	−2.051	29	0.18	0.125	−2.69	−1.883
10	0.19	0.13	−2.95	−2.065	30	0.17	0.12	−2.67	−1.869
11	0.18	0.125	−2.89	−2.023	31	0.17	0.119	−2.67	−1.869
12	0.19	0.133	−2.91	−2.037	32	0.17	0.119	−2.65	−1.855
13	0.18	0.125	−2.87	−2.009	33	0.17	0.119	−2.67	−1.869
14	0.17	0.119	−2.84	−1.988	34	0.17	0.119	−2.67	−1.869
15	0.18	0.125	−2.82	−1.974	35	0.16	0.112	−2.62	−1.834
16	0.18	0.122	−2.8	−1.96	36	0.17	0.119	−2.62	−1.834
17	0.18	0.126	−2.8	−1.96	37	0.17	0.119	−2.62	−1.834
18	0.18	0.126	−2.8	−1.96	38	0.18	0.126	−2.62	−1.834
19	0.18	0.126	−2.8	−1.96	39	0.16	0.125	−2.6	−1.82
20	0.18	0.125	−2.76	−1.932	40	0.17	0.119	−2.6	−1.82
41	0.17	0.119	−2.62	−1.834	46	0.17	0.125	−2.58	−1.806
42	0.16	0.112	−2.6	−1.82	47	0.17	0.125	−2.6	−1.82
43	0.17	0.119	−2.6	−1.82	48	0.16	0.112	−2.58	−1.806
44	0.17	0.119	−2.6	−1.82	49	0.17	0.119	−2.58	−1.806
45	0.17	0.125	−2.6	−1.82	50	0.17	0.119	−2.58	−1.806
51	0.17	0.119	−2.58	−1.806	76	0.15	0.105	−2.45	−1.715
52	0.17	0.119	−2.58	−1.806	77	0.15	0.105	−2.47	−1.729
53	0.15	0.105	−2.58	−1.806	78	0.15	0.105	−2.45	−1.715
54	0.16	0.112	−2.56	−1.792	79	0.15	0.105	−2.47	−1.729
55	0.17	0.119	−2.56	−1.792	80	0.15	0.105	−2.42	−1.694
56	0.16	0.112	−2.54	−1.778	81	0.15	0.105	−2.42	−1.694
57	0.17	0.119	−2.49	−1.743	82	0.15	0.105	−2.42	−1.694
58	0.17	0.11	−2.51	−1.757	83	0.16	0.112	−2.42	−1.694
59	0.17	0.119	−2.51	−1.757	84	0.14	0.098	−2.42	−1.694
60	0.16	0.112	−2.51	−1.757	85	0.15	0.101	−2.42	−1.694
61	0.15	0.105	−2.51	−1.757	86	0.15	0.102	−2.42	−1.694
62	0.16	0.112	−2.47	−1.729	87	0.15	0.102	−2.42	−1.694
63	0.16	0.112	−2.51	−1.757	88	0.14	0.111	−2.42	−1.694
64	0.16	0.112	−2.49	−1.743	89	0.16	0.113	−2.4	−1.68
65	0.15	0.105	−2.49	−1.743	90	0.15	0.105	−2.42	−1.694
66	0.16	0.112	−2.54	−1.778	91	0.15	0.105	−2.42	−1.694

(continued)

Table 1. (*continued*)

N	Max excessive pore water pressure		Cyclic stress ratio		N	Max Excessive pore water pressure		Cyclic stress ratio	
	Without VsD	With VsD	Without VsD	With VsD		Without VsD	With VsD	Without VsD	With VsD
67	0.16	0.112	−2.49	−1.743	92	0.15	0.105	−2.42	−1.694
68	0.16	0.112	−2.47	−1.729	93	0.14	0.098	−2.42	−1.694
69	0.16	0.112	−2.51	−1.757	94	0.15	0.105	−2.4	−1.68
70	0.16	0.112	−2.49	−1.743	95	0.15	0.105	−2.42	−1.694
71	0.16	0.112	−2.49	−1.743	96	0.14	0.098	−2.42	−1.694
72	0.15	0.105	−2.49	−1.743	97	0.15	0.11	−2.4	−1.68
73	0.16	0.112	−2.49	−1.743	98	0.15	0.11	−2.4	−1.68
74	0.15	0.105	−2.42	−1.694	99	0.15	0.105	−2.4	−1.68
75	0.15	0.105	−2.45	−1.715					

(consequently, with complete stress reversal), initial liquefaction may well be produced in each loose and dense specimen. According to the definition this behavior would currently be classified (since the static shear stress was zero) as cyclic mobility. The employment of the term initial liquefaction led several to the error that flow liquefaction can be initiated in any loose or dense cohesion-less soil. The cyclic stress approach is conceptually quite simple; the earthquake induced loading expressed in terms of cyclic shear stresses, is compared with the liquefaction resistance of the soil, and additionally expressed in terms of cyclic shear stresses. At locations wherever the loading exceeds the resistance, liquefaction is anticipated to occur. Application of the cyclic stress approach but, needs careful attention to the way during which the loading conditions and liquefaction resistance are characterized.

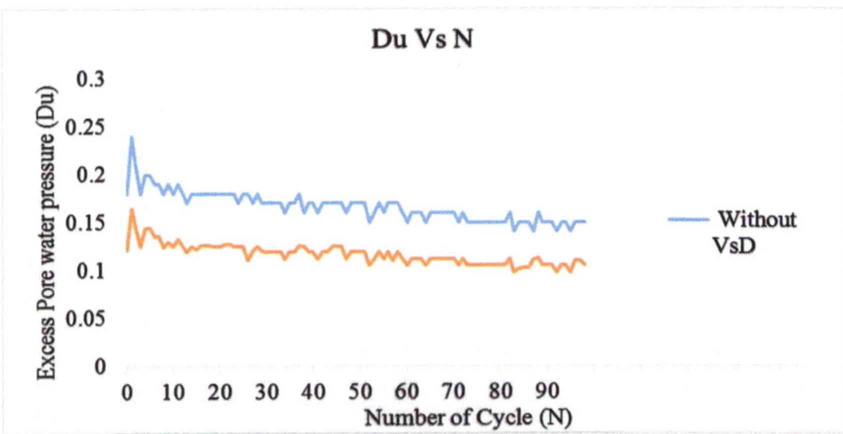

Fig. 2. Dissipation of excess pore pressure with respect to number of cycle

5 Results and Discussion

Table 1 Summarized the value obtained from the cyclic experiments, for both with and without single VsD installation. Graphical representation of Maximum excessive pore pressure (Dn) and Cyclic Stress Ratio (CSR) are shown in Figs. 2 and 3. The results clearly shows that the number of cycles required to develop the pore pressure increases with VsD.

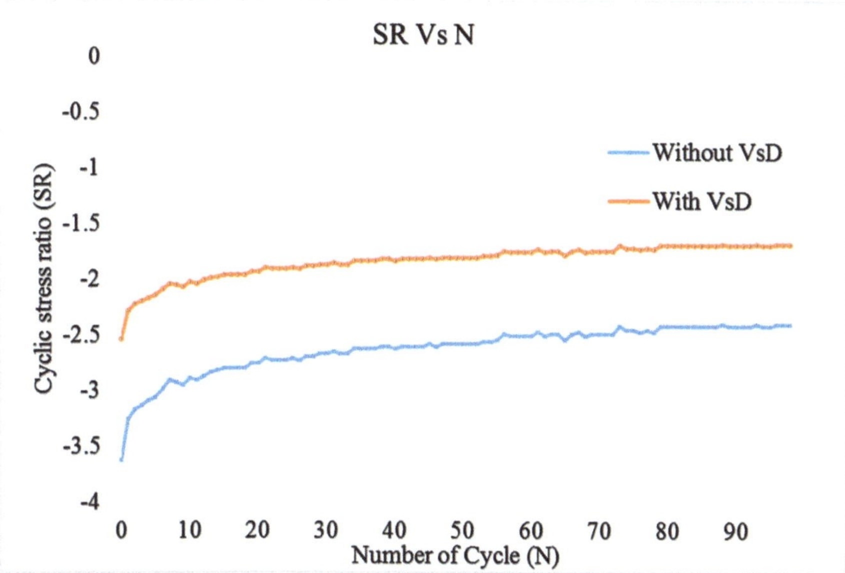

Fig. 3. Development of cyclic stress ratio with respect to number of cycle

6 Conclusion

Based on the present work, the following conclusion are drawn

- The Maximum excessive Pore water pressure is more in initial cyclic loading, at the final number of cyclic loading it decreases in both with and without VsD.
- VsD installed samples results shows nearly 30% reduction in excessive pore pressure dissipation.
- Cyclic Stress Ratio is approximately 30% reduced by application VsD.

Therefore pre-consolidated VsD improved soft clay deposit are better in dynamic behavior, compare to the clay deposit without Sand Drains.

References

Seed, H.B., Idriss, I.M.: Simplified procedure for evaluating soil liquefaction potential. J. Geotech. Engrg. Div. ASCE **97**(9), 1249–1273 (1971)

Kong, X., et al.: Stress-dilatancy relationship of zipingpu gravel under cyclic loading in triaxial stress states. Int. J. Geomecha. ASCE (2016). ISSN:1532-3641

Raj, P.V., et al.: Cyclic and post cyclic behaviour of overconsolidated clays. In: Proceedings of Indian Geotechnical Conference IGC-2014 (2014)

Head, K.H.: Manual of Soil Laboratory Testing. Effective Stress Tests, vol. 3. Wiley, Singapore (2006). (1998)

Anderson, K.H., Pool, S.F., Brown, S.F., Rosenbrand, W.F.: Cyclic andstatic laboratory tests on Drammen clay. J. Geotech. Eng. Div. ASCE **115**(7), 968–983 (1980)

Rakesh, J.Pillai, Nazeeh, K.M., Robinson, R.G.: Post-cyclic behavior of clayey soils. Indian Geotech. J. **44**(1), 39–48 (2013)

Zergoun, M., Vaid, Y.P.: Effective stress response of clay to undrained cyclic loading. Can. Geotech. J. **31**, 714–727 (1994)

Hydo, M., Yasuhara, K., Hirao, K.: Prediction of cyclic behavior in undrained and partially drained cyclic triaxial tests. Soils Found. **32**(4), 117–127 (1994)

Author Index

T. Abdoun and S. Elfass (eds.), *Soil Dynamics and Soil-Structure Interaction for Resilient Infrastructure*, Sustainable Civil Infrastructures, DOI 10.1007/978-3-319-63543-9

MIX
Papier aus verantwortungsvollen Quellen
Paper from responsible sources
FSC® C105338

If you have any concerns about our products,
you can contact us on
ProductSafety@springernature.com

In case Publisher is established outside the EU,
the EU authorized representative is:
Springer Nature Customer Service Center GmbH
Europaplatz 3, 69115 Heidelberg, Germany

Printed by Libri Plureos GmbH
in Hamburg, Germany